Artificial Intelligence in Business and Technology

Accelerate Transformation, Foster Innovation, and Redefine the Future

AD Al-Ghourabi

Notice of Copyright

Edited with help from:

Grammarly

Printed in the United States of America.

First Printing Edition, 2023

ISBN 979-8-9886294-1-2

Dedication

To my parents: I have been blessed with your presence throughout my life. Thanks to your love, guidance, support, and prayers, I am what I am today.

TABLE OF CONTENTS

AUTHOR'S NOTES

Introduction

To characterize 2023 through the lens of scientific and technological advancements, it is undeniable that Artificial Intelligence (AI) will claim the spotlight.

Though AI might appear as a novel concept in our day-to-day lives, in terms of public attention and wider accessibility, it has been a part of theory and practice for decades. There is hardly a computer science curriculum that does not include at least one AI-related subject, and countless scholarly articles and books have been devoted to its theory and possible uses. The narrative has changed recently to the practical use of AI for businesses and our daily tasks.

AI has roots in history, with tales of mechanical beings appearing in Greek mythology. However, only in the mid-20th century did AI become an area for scientific exploration.

The label "Artificial Intelligence" was coined by John McCarthy at a Dartmouth College conference in 1956, signaling AI's birth as a research discipline. In the ensuing decades, AI oscillated between periods of optimism and disillusionment, the so-called "AI winters," as researchers struggled with the complexities of developing intelligent machines (Crevier, 1993).

Nevertheless, seminal breakthroughs in the 1990s and 2000s, particularly machine learning, sparked renewed enthusiasm for AI. The advent of practical applications such as recommendation systems, voice recognition, and image recognition catapulted AI from research labs into the real world (McCorduck, 2019).

By early 2023, AI has become the spotlight, with generative AI tools becoming accessible and highly affordable. It has prompted some technology leaders to propose a six-month mortarium to allow regulations and ethical considerations to catch up.

This has presented a dilemma for me; while the idea of the book has been at the back of my head for a while, with copious notes and materials gathered, the latest developments meant I needed to accelerate this project. While there is a new development every day, I hope to include the latest relevant information by the time the book is published.

About the Book

AI is an invigorated field and is about to transform various facets of our lives and work fundamentally. It is no longer a concept of the future but a practical tool that businesses can now exploit to propel growth, efficiency, and innovation.

This book traces this captivating journey only as a backdrop for AI's progression. From its conceptual origins in antiquity, through the boom and bust cycles of the 20th century, to the remarkable innovations of today, we explore the ever-evolving AI technology.

We look into how AI is transforming traditional business models at a rapid pace, disrupting competitive advantages, and ushering in a new era of digital innovation. With a deep dive into the rise of automation, predictive analytics, and decision-making capabilities, we explore how AI redefines business value creation.

As we stand on the precipice of an AI-driven future, I aim to help readers understand the foundation of this field and help them chart potential courses into the uncharted waters of AI's exciting potential. I hope to enlighten, provoke, and spark readers to ponder rigorously about AI's spectrum of opportunities and dilemmas, arming them with the foundational knowledge needed to exploit its capacities judiciously and ethically.

The advent of AI does not simply mean replacing human jobs with machines. Introducing AI into our systems typically creates new designs with new business models, employment, and workflows. As AI continues to improve, it will create new opportunities and different kinds of organizations. It will shift the dynamic between people and machines and develop new workflows and ways of doing things.

As we consider the future of AI, it is clear that it is not just about developing more sophisticated AI models but also about the societal, economic, and organizational changes needed to realize these technologies' full potential. "The question is not whether AI will be good enough to take on more cognitive tasks but how we will adapt (Agrawal et al., 2022)."

As we journey together, we will highlight the business imperatives and ways businesses can successfully implement AI, using some real-life case studies to bring the concepts to life. Practical examples in healthcare, marketing, sales, financial services, FinTech, education, and IT operations should present a groundswell for ideas to pursue. We will also examine the future of AI and its potential impacts. Above all, we will discuss the ethical considerations that AI brings to the forefront because the responsible use of AI is not just a nice-to-have but a must-have. This book is not about abstract theories or complex equations; it is about the real-world application of AI to drive business value.

As someone who practices and embraces new technology, I could not ignore the latest capabilities in creating this book. I utilized Midjourney to design the book cover and experimented several times before settling on one relevant to the content and audience. ChatGPT was instrumental in helping me adjust the chapters' order and craft the tagline based on the introduction I provided. Bing Chat, now with OpenAI support, helped me as a research tool, almost as a replacement for "Google search." Additionally, Grammarly proved to be an excellent editor and companion, learning my writing style and offering helpful recommendations and suggestions. I have extensively used Grammarly; the tool tells me I was more productive than 99% of its users, analyzing close to 4 million words in total, as I started editing the chapters individually and as the book came together in one coherent product. This experience has given me a great appreciation for the work editors do, but I also marvel at how efficient and productive these tools have become.

AI in Business

AI has transitioned from a trendy catchphrase to a critical discussion point for technology and business leaders. From optimizing processes to crafting personalized customer interactions, AI is fundamentally altering how businesses can function, compete, and, I must add, survive.

The strength of AI lies in its capacity to scrutinize voluminous data sets and extract insights that would be unfeasible for humans to process within a reasonable timeframe. This capability is proving powerful in a world increasingly fueled by data. Whether forecasting consumer behavior, identifying deceptive transactions, or enhancing supply chains, AI is revolutionizing all aspects of the business.

Nevertheless, the influence of AI extends beyond boosting a business's operational efficiency. It is simultaneously reshaping business architectures, spawning novel revenue channels, and resetting the rules of competitive engagement. Enterprises that adopt and harness the power of AI are poised to be tomorrow's frontrunners—those who falter risk disruption.

Lastly, an astute reader may note that these are data science functions. That would be a correct observation, mainly because these two fields are inherently intertwined and feed off each other. On the one hand, data science employs statistical tools and programming languages to interpret, analyze, and extract valuable insights from vast datasets. These insights are pivotal in influencing business strategies and decision-making processes. On the other hand, AI, specifically machine learning, utilizes these data insights to train algorithms, enabling them to learn and make intelligent predictions or decisions without explicit programming. The two fields coexist harmoniously, each enhancing and empowering the other. Thus, understanding data science equips you with the fundamental knowledge to explore further into the AI field and vice versa. That is why concepts, techniques, and methodologies often seem interchangeable and blurry around the edges. This overlap is not a bug, but a feature, encouraging cross-disciplinary learning, innovation, and progress in these influential domains of the technology world.

Intended Audience

This book is intended for business and technology leaders and decision-makers keen to grasp AI's true implications for their organizations. Whether you are a CEO eager to appreciate the strategic connotations of AI, a CTO entrusted with its implementation, a government official exploring enacting regulations for AI, or a team leader aspiring to capitalize on AI in your workloads, this book will provide you with the requisite knowledge and insights.

The book should also prove helpful for individuals lacking a technical pedigree but eager to comprehend AI beyond the buzz. A technical degree is optional to understanding this book. All it demands is a spark of curiosity and a readiness to explore.

The book's content may appeal to a broad audience, so some chapters may contain redundant information with gradual progression. The goal is to elaborate on ideas while allowing readers to skip to sections that interest them the most. Also, topics like ethics will be covered from different angles related to the topic being discussed.

Style-wise, except for the chapter on implementing AI in your business which was more prescriptive in nature, my writing was terse and to the point, an approach I learned during my Executive MBA program at the University of Denver that I found very helpful in synthesizing and articulating relevant information.

Lastly, although I have taught graduate-level courses, spoken at conferences, and written a few publications, writing a book is a different endeavor altogether. As this is my first published book in the business and technology domain, your feedback is welcome and highly appreciated.

Chapter 1

HISTORY AND FOUNDATION

Artificial intelligence (AI) has introduced a paradigm shift in how technology is perceived and integrated into our daily lives (Frąckiewicz, 2023). We are probably witnessing the fourth industrial revolution, with the convergence of the physical and digital worlds. Data science and AI are leading this transformation, disrupting industries, creating new business models, and redefining customer expectations. AI is slowly ingraining itself into every facet of our lives, unbeknownst to us, from healthcare to financial services, manufacturing to education.

Evolution and History of AI

AI is not a novel concept, and its roots date back to the middle of the 20th century when British mathematician and logician Alan Turing pondered the possibilities of machines that could mimic human intelligence. This quest laid the foundation for the AI capabilities we have today. Over the decades, AI has seen significant developments and milestones shaping today's technology (Gold, 2023).

AI, in its myriad forms, has been a source of fascination and inspiration for humanity throughout the ages. Its earliest roots can be traced back to ancient cultures, where mythologies were replete with stories of automatons - mechanical devices that mimicked human actions. Over time, humanity's fascination with replicating its intelligence evolved from these rudimentary mechanical devices to today's sophisticated and far-reaching AI technologies.

Philosophical Foundations and Early Ideas

The concept of AI can trace its roots back to ancient civilizations. Greek myths of Hephaestus, the blacksmith who manufactured automated servants, and the idea of the "Golem" in Jewish folklore, bring to life the old-fashioned fascination with creating artificial life (Shashkevich-Stanford, 2019; Oreck, 2008). In the 17th and 18th centuries, philosophers such as René Descartes and Gottfried Leibniz posited ideas about thinking machines and calculative logic (McCorduck, 2019). The idea of creating artificial life or intelligence has been an enduring theme throughout history. They advanced early ideas that planted seeds for our current understanding of technology and computation. Descartes, notably, conceptualized the universe as a mechanical system, presenting humans and animals as complex machinery. Alternatively, Leibniz, creator of the calculating machine, fostered a universal language of symbols and numbers, a concept bearing remarkable similarity to today's programming languages.

With the advent of modern computing, British mathematician Alan Turing proposed the universal Turing machine during World War II, a theoretical model capable of executing any calculation given sufficient time and memory resources (Henderson, 2007). This conceptual leap provided a crucial foundation for contemporary computer science and, by extension, AI.

In 1950, Turing made another significant stride in AI. His seminal paper, "Computing Machinery and Intelligence," put forward the concept of machine learning, suggesting that machines could be designed to replicate human learning patterns, thereby enhancing their performance incrementally (Turing & Copeland, 2013).

Turing also introduced the "Turing Test" in the same publication that aims to ascertain a machine's ability to demonstrate intelligent behavior (Turing & Copeland, 2013). If a machine could produce responses in a conversation indistinguishable from a human's, it would pass the Turing Test, demonstrating intelligence (Henderson, 2007). Against the backdrop of modern computing's birth and early AI, Turing's innovations represent critical milestones.

Turing's theoretical work has had profound implications for the development of modern AI, influencing everything from machine learning algorithms to AI's role in various industries today.

Dartmouth Conference and the Birth of AI as a Discipline

The Dartmouth Conference convened in 1956, is frequently hailed as the genesis of AI as an autonomous field of study (Gold, 2023). Visionaries like Marvin Minsky, John McCarthy, Allen Newell, and Herbert Simon played pivotal roles. The groundbreaking hypothesis that "every aspect of learning or any other characteristic of intelligence can in principle be so accurately delineated that a machine can be crafted to mimic it" served as one of the cornerstone assertions of this conference and has significantly guided the course of AI research and evolution (Crevier, 1993). The conference marked the beginning of an era where the pursuit of creating machines capable of simulating human intelligence became a recognized field of study.

The Dartmouth Conference anticipated that machines would reach human-level intelligence within a generation, which may prove accurate. This bold prediction has been a source of inspiration and debate within AI research for decades. However, the tools, concepts, and methodologies developed in the pursuit of this goal have profoundly impacted not just technology but also various other sectors, including the economy, where AI has the potential to reshape job markets and create a broad range of services.

Early Cycles of Boom and Bust

The inception of AI as a formal academic discipline in 1956 marked the beginning of what is often termed the "golden age" of AI (1956-1974) (Nilsson, 2010). This period saw an influx of government and private funding, resulting in significant strides in AI research (Henderson, 2007). Early AI programs, like Samuel's checkers' program, Newell and Simon's Logic Theorist, and McCarthy's Lisp language, held the promise of machines that could mimic human intelligence.

However, the progress slowed as the AI research community realized the complexity of problems in language understanding, learning, and commonsense reasoning. These issues, coupled with the failure to meet the overly ambitious expectations set by the AI community, led to "disillusionment" and the onset of the first "AI winter" in the mid-1970s, a reference to alternating periods of optimism and disappointment and a period of reduced funding and interest in AI research (Crevier, 1993).

Following this was a period of resurgence in the 1980s, often called the "second summer" of AI. This era was marked by the advent of "expert systems." These were AI programs designed to provide solutions in specific domains, such as medical diagnosis or geological exploration, by mimicking the decision-making ability of a human expert. These systems, like MYCIN and Dendral, stirred a new wave of optimism and investment in AI (Russell, 2016).

The expert systems, though a breakthrough, were not without their limitations. They were expensive to build and maintain, relied heavily on the knowledge of human experts, and failed to deliver broad-based solutions. The market for expert systems could not sustain the hype, and by the end of the 1980s, AI plunged into its second winter.

Unfolding of Contemporary AI

The late 1990s and early 2000s marked the beginning of the modern era of AI, characterized by amplified computational capabilities, accessibility of expansive datasets, and progress in learning algorithms. Machine learning, later evolving into deep understanding, began to eclipse conventional AI methodologies.

A pivotal event during this period was the victory of IBM's Deep Blue over the incumbent world chess champion, Garry Kasparov, in 1997 (Henderson, 2007). Deep Blue, while not embodying AI as we comprehend it today, primarily relied on exhaustive search techniques rather than learning from data. Nonetheless, its triumph marked an impressive achievement in computer science and showcased the possible capabilities of machines in tasks conventionally viewed solely within the realm of human intellect.

This event's significance reverberated beyond chess, influencing public perception, and igniting renewed interest in AI. Moreover, it spurred further research into developing algorithms that could learn and adapt rather than relying solely on brute-force computations.

The emergence of the Internet during the closing years of the 1990s and the early years of the new millennium substantially impacted the course of contemporary AI. The Internet served as an accelerant, triggering an exponential surge in the availability of digital data. The deluge of data and enhancements in data storage and processing technologies laid the groundwork for AI's rapid growth.

This period witnessed several significant milestones that breathed new life into AI research and applications, steering it towards the path of modern AI we know today. The emergence of machine learning marked a shift from rule-based AI systems, which relied on hardcoded knowledge, to systems that could learn from data. Machine learning algorithms could identify patterns and make predictions based on data, negating the need for explicit programming for each task. This was a significant step forward, offering a more flexible and scalable approach to tackling complex problems.

AI manifested slowly in diverse facets of the Internet, from search engines to digital advertising. Google, a scholarly endeavor initiated by Larry Page and Sergey Brin, transformed the web search experience by introducing its PageRank algorithm.

AI techniques were used to rank web pages based on relevance and quality, drastically improving search results.

Similarly, recommender systems emerged as a powerful application of AI, driven by the need to navigate the vast digital world. Companies like Amazon and Netflix use AI to analyze user behavior and preferences, providing personalized recommendations that enhance user engagement and satisfaction (Barazy, 2023).

Another area that saw a surge in AI applications was online advertising. Platforms like Google AdWords and Facebook Ads use AI algorithms to target advertisements based on user data, enhancing the effectiveness of online advertising campaigns.

The concluding years of the 20th century and the initial years of the 21st laid the cornerstone for contemporary AI, leading to further developments in machine learning, deep learning, and AI applications. A confluence of high computational capabilities, data accessibility, and advancements in AI algorithms distinguishes our era and ushers in a new disruptive technology tantamount to the spread of internet accessibility and use.

The early cycles of boom and bust are not uncommon in the journey of disruptive technologies. The "hype cycle," a concept introduced by Gartner, visually represents specific technologies' maturity, adoption, and social application (Barazy, 2023). It is characterized by the "peak of inflated expectations," followed by the "trough of disillusionment," the "slope of enlightenment," and finally, the "plateau of productivity (Smith, 2006)." AI's journey resembles this model, navigating the peaks and troughs of expectations and results (Wiles & Jaffri).

AI is undoubtedly in a prolonged "summer" phase, backed by significant advancements in machine learning and deep learning, big data availability, and computational power improvements. Optimism is stronger than ever, but the journey has taught us to be cautious of the hype. As we push the boundaries of AI, it is vital to balance optimism with realism, ensuring that expectations align with the current capabilities of AI.

To truly understand AI's present and future implications, it is essential to grasp its underlying mechanics and structures. In the next section, we examine the foundations of AI and explore the different disciplines that underpin this revolutionary technology.

Foundations of AI

Machine learning and deep learning are the foundation of AI. The study of AI involves the computations that enable perception, reasoning, and action. AI has grown unprecedentedly over the past 25 years, with robots now performing human-like tasks. Sophia, an AI-based humanoid, was awarded Saudi Arabian citizenship and interviewed world leaders without manual intervention (Abbass, 2022). In early May 2023, an unprecedented AI-generated debate united two iconic figures: Socrates and Bill Gates (AI Tech Talk, 2023). This thought-provoking exchange, collapsing more than 1500 years in time disparity, explored various crucial topics, including technology, ethics, and AI's future implications and risks.

Sub-Domains of AI

AI is an offshoot of computer science committed to engineering machines capable of emulating intelligent behavior. "Intelligence" pertains to the capacity to learn, reason, troubleshoot, perceive, and even utilize language. As time has passed, AI has grown to encompass an array of specialized domains, each with distinct aims and methodologies. However, machine learning and deep learning remain the bedrock of AI.

As a field, AI encompasses numerous sub-domains that collectively characterize intelligent machines.

Machine Learning, the foundation of AI, equips machines to learn from data and make informed predictions without explicit programming (Kumar, 2023). It forms the foundation for various applications and advancements, substantially enhancing decision-making and enabling automated systems to thrive. Instead of coding instructions for a task, a machine learning model is fed examples, data, and assimilating patterns. This training enables it to generate predictions or decisions when presented with new data.

Deep learning, a subdivision of machine learning, is modeled on the human brain's structure. It leverages artificial neural networks with multiple layers—hence, the descriptor "deep" ("Introduction to Machine Learning," 2021). Deep learning models excel at deciphering complex patterns in substantial data sets, making them particularly adept at image and speech recognition tasks.

Natural Language Processing (NLP) is about the understanding and processing of human language by machines, empowering AI systems to interpret, translate, and interact with us meaningfully (Barazy, 2023). From language translation to sentiment analysis and chatbot interactions, NLP bridges the gap between human communication and machine comprehension. One key component empowering NLPs is data tokenization, a process by which a body of text is divided into sub-units, or tokens. Tokens can be words (e.g., smarter), characters (e.g., s-m-a-r-t-e-r), or even sub-words (e.g., smart and er) (Pai, 2020).

Computer Vision empowers machines to perceive and interpret visual information, enabling them to analyze images and videos, recognize objects and faces, and facilitate advancements in areas such as autonomous vehicles and surveillance systems ("Introduction to Machine Learning," 2021). Through sophisticated algorithms, machines gain the ability to navigate and understand the visual world.

Robotics combines the power of AI with physical systems, resulting in machines that interact and operate in the physical world with astounding capabilities of understanding their surroundings and creating real-world perceptions. They navigate with motion planning, seamlessly interacting with humans, marching toward a revolutionized future of unprecedented synergy between humans and AI-driven systems.

Reinforcement Learning introduces the concept of training AI agents to make sequential decisions through feedback from their environment. This sub-domain enables machines to excel in game playing and autonomous control, paving the way for intelligent systems that adapt and improve over time.

Expert Systems replicate the knowledge and reasoning of human experts within specific domains, providing specialized advice and solutions (Gold, 2023). These AI systems augment human expertise, enabling effective decision-making and problem-solving in complex scenarios.

Knowledge Representation and Reasoning focus on structuring and organizing information to facilitate reasoning and inference. By developing formal frameworks for representing knowledge and enabling machines to draw logical conclusions, this sub-domain empowers AI systems to make informed judgments.

Planning and Scheduling address the complexities of generating optimal plans and schedules for complex tasks and activities. AI systems navigate complex scenarios through advanced algorithms and techniques, optimizing resource allocation and enhancing industry efficiency.

Machine Vision, different from "computer vision" discussed earlier and used in industrial machinery, explores extracting meaningful information from visual data, including image processing, object recognition, and quality control in manufacturing. With the ability to analyze visual content, machines play a pivotal role in enhancing accuracy and precision in various applications.

Natural Language Generation (NLG), also known today as Generative AI, enables machines to produce coherent and meaningful written or spoken content, mirroring human-like language. From generating reports to crafting compelling narratives, NLG facilitates communication and creates possibilities for innovative content creation.

Given the meteoric rise of NLG and its broader use in productivity tools, the following section covers the topic in more detail.

Generative AI

Generative AI is an intriguing branch of AI that deals with creativity and imagination. At its core, generative AI employs complex algorithms and models to produce new data that closely mimics the patterns and characteristics of the input it has been fed. Adobe's Generative Fill, for example, is an innovative tool in Photoshop that uses AI to create and modify images based on simple text prompts (Johnson, 2023). Under the covers, it is driven by Adobe Firefly, a set of AI models trained on diverse visual content. This tool lets users quickly and efficiently bring their vision to life, offering significant control over the creative process, whether adding new elements, altering existing ones, or creating an image from scratch. Generative Fill is not just about creation - it also allows for efficient refinement and enhancement of images, such as adjusting outfits, switching backgrounds, or adding objects based on a text description. Generative Fill also respects the integrity of the original image. It adds to pictures in a way that blends seamlessly with existing elements, considering factors like shadows, reflections, and perspective. If a user wishes to remove an element, it intelligently fills in the gap with content that fits the context of the image (Johnson, 2023).

In essence, Adobe's Generative Fill combines the power of AI with human creativity, offering a powerful tool for graphic designers. It is a testament to how AI can amplify human potential and open new avenues for creativity.

Generative AI applications can be far-reaching with limitless creativity. It is reasonable to imagine machines composing symphonies rivaling classical composers' masterpieces or painting breathtaking artworks that could be hung on the walls of prestigious galleries with vibrant characters and captivating storylines. In April 2023, an AI-generated record track for the rapper Drake went viral, rivaling some of his prior hits, and was indistinguishable from the real Drake, prompting a debate on copyright infringement and content authenticity (Li, 2023).

Generative AI can revolutionize the fashion, design, and advertising industries. Generating fresh and innovative concepts pushes the boundaries of what was once possible. It infuses novelty into our lives and opens avenues for unexplored possibilities in a democratized fashion. It stands as a testament to the limitless potential of human ingenuity combined with the computational prowess of AI.

Chapter 2

RESHAPING BUSINESS AND INDUSTRIES

Artificial Intelligence (AI) is more than another technology businesses can adopt or ignore. It is a transformative power that can fundamentally alter competition by redefining industries and modifying traditional business models. This paradigm shift can create entirely new systems, business models, jobs, and workflows, thus underscoring the need for business and technology leaders to understand and adapt to these changes.

Technology advancements have consistently influenced how companies function and compete, and AI is no different. However, the impact of AI is more comprehensive than operational effectiveness or productivity enhancements alone. It signifies a shift that gives rise to new prospects for business value creation and competitive edge and disrupts current business models in an unprecedented fashion.

Altered Competitive Environment

As AI technology evolves rapidly, business leaders must prepare for its transformative impact. AI offers immense business potential and requires thoughtful leadership to address the opportunities and challenges it poses proactively. The goal is to navigate organizations toward a future where AI is a powerful catalyst for growth, innovation, and positive social impact.

Understanding AI's implications on innovation and competition is vital. In a future AI-dominated world, innovation becomes a survival imperative rather than an option. AI is not merely improving existing products and services but enabling entirely new ones, disrupting traditional markets, and birthing new business models. From healthcare's AI-powered diagnostics to finance's AI-based predictive analytics and automated trading systems, AI's revolutionizing influence is pervasive. To remain competitive, business leaders must closely monitor these dynamics, invest in AI-driven innovation, and cultivate a continuous learning and agility culture. Companies must become comfortable with the idea of self-disruption.

Workforce transformation is another profound impact of AI. Automation is reshaping job opportunities, with expertise in AI and machine learning becoming increasingly sought after. This transition presents challenges and opportunities for businesses to enhance productivity and foster more creative and fulfilling work for employees. Effective transition management is crucial, possibly involving retraining employees, redesigning jobs, or adopting new hiring practices. Leaders must also commit to equitable practices, ensuring the benefits of AI, such as increased productivity, higher wages, and improved work-life balance, are distributed broadly.

While AI may replace some traditional roles, it will create new jobs and industries and help upskill the broader population. Generative language models, for example, enable more people to write well, design graphics, and realize ideas without much training, further playing the level field and potentially increasing competition.

AI's contribution to ESG initiatives can significantly bolster a company's reputation and profitability. By optimizing resource use, minimizing waste, and improving energy efficiency, AI can help businesses demonstrate their commitment to environmental sustainability. This resonates with customers, stakeholders, and investors, who increasingly value corporate social responsibility, enhancing the company's reputation.

Moreover, these sustainable practices can directly benefit the bottom line. Efficient resource utilization and waste reduction can lead to substantial cost savings. Similarly, improved energy efficiency can result in lower operating expenses. AI can also help optimize logistics, leading to more cost-effective and environmentally friendly operations. Thus, AI's ability to aid in realizing ESG initiatives serves global sustainability goals and helps companies gain a competitive edge and achieve economic efficiency. It is, therefore, imperative for executives and boards to consider AI as a critical component of their ESG efforts.

AI's Impact on Different Industries

AI presents tremendous opportunities spanning industries and functions. They can be seen in marketing, sales, customer service, operations, human resources, and finance, among others. AI can analyze customer data in marketing and sales to provide personalized recommendations, predict future buying behavior, and optimize marketing campaigns. It also automates tasks such as customer segmentation and lead scoring.

AI-powered chatbots and virtual assistants are revolutionizing customer service by providing 24/7 support, handling multiple queries, and predicting customers' needs before they arise (Mustapha, 2023). In modern organizations, IT operations are evolving. The author coined the term "Intelligent Ops," an intersection of IT, security, development, and business operations. "Intelligent operations allow organizations to elevate business performance and advance value creation through predictability, automation, and actionable insights....using Big Data analytics, machine learning, and automation (Al-Ghourabi, 2022)."

AI's role in human resources and finance is also transformative. It streamlines recruitment by screening resumes, scheduling interviews, predicting job fit, personalizing employee training, and predicting attrition. In finance, AI automates processes like invoice processing and expense management, detects financial fraud, and assists with investment decisions by analyzing financial markets (Mustapha, 2023).

Beyond these functions, AI is revolutionizing multiple industries, such as healthcare, manufacturing, retail, and transportation (Mustapha, 2023). In healthcare, AI assists in diagnostics, plays a crucial role in drug discovery, and enables personalized treatment plans. In manufacturing, AI transforms processes by optimizing production lines, enabling predictive maintenance, and driving demand forecasting.

Retail is another sector where AI enhances the customer experience and optimizes operations through recommendation engines, dynamic pricing strategies, and streamlining inventory management (Al-Ghourabi, 2023). AI is innovating through self-driving technology and optimizing logistics operations in the transportation sector (Mustapha, 2023).

AI is also revolutionizing product differentiation and competitive advantage. Companies traditionally relied on broad market segments for creating and marketing their products. However, AI allows them to go deeper, enabling companies to tailor their products and services by analyzing consumer preferences to meet specific customer needs. This real-time personalization increases customer satisfaction, loyalty, and the company's bottom line.

Operational efficiency is another area where AI is excelling. AI applications in supply chain management have significantly improved operational processes, predicting demand, optimizing inventory, identifying bottlenecks, automating logistics processes, improving quality control by detecting defects in products or procedures, and suggesting the most efficient delivery routes. It also helps reduce operating costs and waste, thus bolstering profitability and freeing business leaders to focus more on value creation.

Risk management and information security, too, are seeing AI's influence. AI can predict potential threats, detect anomalies indicating fraud or security breaches, and assist enterprises in regulatory compliance.

AI's broad impact continuously evolves, extending to various operations, driving efficiency and cost savings, creating new jobs, and reshaping entire industries. It automates routine tasks, freeing employees to "focus on more complex and creative aspects of their work ('Promising Future of AI-Enabled Web Development,' 2023)."

These examples illustrate AI's revolutionary role across various industries by driving innovation, transforming business models, improving operational efficiency, enhancing customer experiences, and creating new growth opportunities. As AI advances, its impact across industries will continue to grow and evolve, driving a virtuous cycle of further innovation and transformation. In later chapters, we will cover these topics in a lot more detail.

Real-World Success Stories

One of the most comprehensive surveys of AI adoption in businesses was conducted by Forbes Advisor, who surveyed 600 business owners using or planning to use AI in their companies. According to their findings, AI is used in many areas, from customer service and fraud management to customer relationship management and digital personal assistants.

Numerous businesses have harnessed AI to create value. For example, Netflix uses AI to personalize recommendations to its hundreds of millions of users. This has been a critical factor in its growth and success. American Express utilizes AI to detect real-time fraudulent transactions, saving millions annually. Coca-Cola uses AI to analyze social media data and understand consumer preferences, aiding in creating new products and marketing strategies. IBM leverages its AI platform, Watson, to simplify internal processes, enhance decision-making, and innovate new services.

AI has been significantly influential in the FinTech sector. Kensho, a part of S&P Global, "combines natural language search queries, graphical user interfaces, and secure cloud computing to create a new class of analytics tools for investment professionals. Kensho addresses the three biggest challenges surrounding investment analysis on Wall Street: speed, scale, and automation. Kensho's data analytics and AI can answer complex financial questions in plain English ('A Hub for Innovation and Transformation,' 2023)."

Enova is an online financial services provider headquartered in Chicago. During the pandemic, its simple online application process has helped more than 7 million individuals handle financial emergencies and access trustworthy credit (Jacob, 2021).

AI has shown immense potential in the education sector to revolutionize learning. For instance, AI-powered platforms like Knewton and Coursera are personalizing the learning experience by creating customized lesson plans and providing targeted feedback. Automated grading systems like Gradescope free up teacher time, while AI tutoring systems like Carnegie Learning's Cognitive Tutor provide immediate, personalized instruction. Adaptive testing platforms adjust the difficulty level based on student performance, and virtual assistants like Amazon's Alexa are making learning more interactive. Furthermore, AI's role in gamification, as seen in "DragonBox Algebra 5+," adds an engaging dimension to learning. All these examples demonstrate AI's transformative role in making education more efficient, effective, and personalized (Karandish, 2021).

At an international level, AI has demonstrated its value as a crucial tool in addressing global problems, notably during the COVID-19 pandemic. For instance, AI has been essential in protecting public health by tracing the virus's trajectory, simulating potential variant spread, and facilitating new treatment development (Gudigar et al., 2021).

AI has seen the most comprehensive penetration and adoption in content creation and curation. Generative AI tools, such as Bing Search, ChatGPT, Forefront, and others, can generate content, ranging from news articles to personalized marketing materials, and curate content to provide customized experiences for users. Moreover, prompt completion has enabled virtually everyone to access unparalleled ingenuity and creativity. This has given rise to new business opportunities and models in the content industry. WriteSonic, for example, is a subscription-based solution that allows generating content in bulk with few keywords and clicks. Grammarly has augmented its editorial capabilities with prompts that help rephrase or re-tone content for better readability and flow of content.

Use Case: Azure OpenAI

Azure OpenAI Service, for example, "is a platform that allows businesses to leverage OpenAI's powerful language models, such as GPT-3 and Codex, but fine-tune them to better suit their specific needs (Al-Ghourabi, 2023)." The platform provides an array of models that companies can use without starting from scratch, enhancing efficiency and effectiveness in achieving accurate results (Wiggers, 2022). Some models have been optimized for data extraction, while others provide varying customizable degrees of imagination. You can use existing prompts or build your own and generate code in a plethora of languages that can integrate with Azure. This offering, for example, suddenly empowers developers to integrate AI/ML capabilities within their applications, short-circuiting the long path of data science teams and expensive models (Wiggers, 2022).

"One of the key improvements in Azure OpenAI Service comes from the introduction of InstructGPT, a family of GPT-3-based models designed to generate less problematic text and align more closely with a user's intent (Al-Ghourabi, 2023)." This has resulted in significant improvements in text generation, making the system more reliable and user-friendly (Wiggers, 2022). Additionally, Azure OpenAI Service includes a new collection of models called "embedding models," which are designed to excel at tasks such as text similarity, text search, and code search, further expanding the platform's versatility (Wiggers, 2022). Moreover, Azure OpenAI Service includes access to Codex, a model capable of generating code given a natural language prompt. This means that users can input a task or problem in plain English and Codex will generate the corresponding code in various languages, including Python, JavaScript, Go, Perl, PHP, Ruby, Swift, and TypeScript. This can be used for tasks like transpilation, explaining code, and refactoring code (Wiggers, 2022).

The service also includes a "responsible AI" system designed to filter out content related to sex, violence, hate, and self-harm, highlighting Microsoft's commitment to ethical AI practices. This system attempts to detect patterns of abuse or harm and, if identified, prompts a dedicated team to investigate and block any misuse (Wiggers, 2022).

Companies have already found innovative applications for the Azure OpenAI Service. For instance, CarMax used the models to sort through and summarize car reviews. At the same time, Farmlands, a New Zealand-based rural supplies cooperative, utilized the service to summarize and classify customer interactions (Hopwood, 2022).

AI's Impact on Information Technology

AI's influence on technology is far-reaching and transformative, permeating multiple sectors. Its effects, from production lines to consumer experiences, are profound as the shift occurs toward a more digitized and interconnected world.

In automation, AI has amplified the capabilities of robots, pushing past the limitations of conventional programmed machines. Robots equipped with AI can now respond to a broader array of challenges and situations. They can adapt to fluctuations in their environment, allowing them to carry out complex tasks that once required human intervention. A prime example is the warehouse industry, where AI-powered robots perform tasks such as inventory management and restocking shelves, demonstrating significant operational efficiency and accuracy strides.

AI's impact extends to improving the intuitiveness of digital interfaces. This results in seamless human-technology interactions and bridges the knowledge gap for users. AI technologies like OpenAI's ChatGPT have transformed from tools to responsive learning companions. These virtual assistants interpret and respond to user commands, thus revolutionizing how we interact with technology. Companies like Microsoft are integrating AI capabilities within their product suites to capitalize on these advancements. For instance, the launch of Copilot as an add-on to their O365 suite demonstrates the importance of integrating generative AI technologies into the daily workflow, fostering a more collaborative and efficient work environment and further individual empowerment (Spataro, 2023).

The cybersecurity field has also greatly benefited from AI's capabilities. AI's potential to identify patterns indicative of cyber threats has given rise to proactive defense strategies (Mustapha, 2023). This shift toward preemptive actions significantly reduces the risk of data breaches and fortifies the protection of digital assets. As of 2023, AI tools have become a bulwark in cybersecurity, aiding in data protection and creating opportunities for novel career paths and industry growth (Raviv, 2021).

AI can help developers write code more efficiently in software development by suggesting code snippets and detecting errors in real-time. This can significantly reduce the time and effort required to develop and maintain software applications.

In website design and development, AI can help designers create more personalized and engaging user experiences by analyzing user behavior, navigation paths, and preferences. This can lead to higher conversion rates and improved customer satisfaction ("Promising Future of AI-Enabled Web Development," 2023). In IT operations, AI can help automate routine tasks such as monitoring system performance and detecting anomalies. This can free up IT staff to focus on more strategic tasks and improve the overall efficiency of IT operations.

In quality assurance, AI can help write testing plans, automate testing processes, and identify defects more quickly and accurately, reducing the time and cost required to ensure software applications meet the desired quality standards. In incident response, AI can help "detect and respond to security incidents more quickly and effectively (Keyser, 2023)." This can reduce the impact of security breaches and improve the overall security posture of an organization.

In customer service and support, AI can help automate routine tasks such as responding to customer inquiries and processing straightforward service requests. This can improve the efficiency of service operations and enhance the customer experience.

In data analytics, AI can process and analyze vast amounts of data at unprecedented speed and accuracy. This enables organizations to derive valuable insights and make informed decisions. AI-powered data analytics tools can identify patterns and correlations that would otherwise be unrecognized, providing a competitive edge to businesses.

In network management, AI can monitor and optimize network performance in real-time. This improves network reliability and reduces downtime. AI can proactively predict and prevent network failures before they occur, ensuring seamless connectivity. In cloud computing, AI can optimize resource utilization and reduce costs, improving the efficiency and scalability of cloud-based applications. In project management, AI can provide real-time insights and recommendations to project managers, improving project outcomes and reducing the risk of project failure. AI-powered project management tools can streamline workflows and enhance collaboration, increasing productivity.

Looking into the future, the ever-accelerating pace of technological evolution demands adaptability from AI technologies. Given the rapid advancements in machine learning and AI methodologies, it is pivotal that these technologies can evolve and acclimate to the changing technological environment. This will ensure that AI remains a relevant and impactful force, continuing to reshape technology and, by extension, our society and economy.

Overall, AI plays a transformative role in multiple areas of IT, driving innovation and improving efficiency across the board. As AI technology evolves, its impact on IT will likely grow even further.

Chapter 3

BUILDING AN AI-READY CULTURE

An AI-ready culture is an organizational atmosphere that embraces artificial intelligence at all levels, understands its importance, and actively promotes its use to improve business operations. While having advanced AI technology and technical skills are prerequisites, the culture promotes these elements as a catalyst for innovation and advancement (Morace, 2014).

An AI-ready culture thrives on curiosity, agility, adaptability, and a continuous desire to learn, improve, and advance. Its core values include data-driven decision-making, embracing innovation, the willingness to learn from failure, and an informed understanding of the ethical implications of AI. These values act as the guiding principles that direct organizational actions and behaviors. Data-driven decision-making means that every decision is made with solid evidence, not intuition or opinion. Embracing innovation enables out-of-the-box thinking and encourages experimentation with creative and disruptive ideas (Bruhn, 2022).

The willingness to learn from failures, rather than avoiding them, fosters resilience and adaptability and promotes a culture of continuous learning. It allows the organization to grow and become more robust with every setback.

Lastly, understanding the ethical implications of AI is equally important to ensure the responsible use of technology. This includes ethical data usage and transparency in AI systems and taking steps to avoid and mitigate bias in AI models.

Fostering Data Literacy

Data literacy, the ability to interpret, analyze, and understand data, is the cornerstone of an AI-ready culture. An organization's ability to create knowledge and insights out of the data it collects and generates is a competitive advantage in a world increasingly driven by data. In the age of AI, data literacy should be a universal skill shared and understood across all levels of the organization.

Promoting data literacy across the organization requires systematic planning and implementation. This can be accomplished through various methods, including training programs, workshops, online learning platforms, and nurturing a data mindset, where every decision and strategy is backed by data, fostering a culture of evidence-based decision-making.

It is important to note that data literacy is about more than just turning everyone into data scientists. Instead, it is about ensuring that all employees have a fundamental understanding of data - how to interpret it, how it can be used for decision-making, and how to appreciate its value. This empowers everyone in the organization to contribute to a culture ready to embrace AI.

Encouraging Experimentation and Learning

Fostering a culture that encourages experimentation and continuous learning is vital to becoming AI-ready. Experimentation is the key to innovation and discovery. It involves taking calculated risks, venturing into uncharted territories, and learning from the outcomes. Experimentation demystifies AI, making it less of a black box and more of a tool that can be understood and manipulated for the organization's benefit.

Creating a safe environment for experimentation is the first step toward promoting a culture of innovation. This means ensuring employees are not penalized for failure but encouraged to learn from it. It is about creating an environment where curiosity is rewarded, and innovative ideas are celebrated (Ammanath, 2022).

Promoting a culture of experimentation and learning requires leadership buy-in, which is instrumental in establishing the right mindset and providing the necessary resources. Leaders should encourage their organizations to experiment with AI tools, ask questions, make mistakes, and learn from them. This enhances their understanding of AI and fosters a culture of innovation and continuous learning.

Ethical Considerations and Responsibility

AI ethics is a complex and crucial part of building an AI-ready culture. It involves a deep understanding of AI's implications, particularly regarding data privacy and bias. As organizations venture into the world of AI, they must grapple with these ethical questions and ensure their use of AI aligns with the principles of fairness, accountability, and transparency.

The role of ethics in data privacy is twofold. First, organizations must respect privacy regulations and guidelines when collecting, storing, and using data. Second, they must also provide that the AI systems they develop respect individual privacy. This requires transparency about how AI systems use data and make their decisions.

Bias in AI is another significant ethical consideration. AI systems learn from data; if the data they learn from is biased, the AI system will also be limited. It is, therefore, crucial to ensure that the data used to train AI systems is as unbiased as possible. To ensure fairness, it is essential to comprehend the origins of bias, devise methods to alleviate it, and consistently scrutinize AI systems for any signs of bias.

AI Adoption and Roadmap Development

Adopting AI in an organization's operations takes time; it requires careful planning and assessment. Before diving into AI implementation, organizations should thoroughly assess their readiness for AI. This includes evaluating their existing technological infrastructure, data management practices, and the skills and capabilities of their workforce.

Organizations must define clear AI objectives and priorities aligning with their business goals. This involves understanding what they hope to achieve with AI, the problems they aim to solve, and how AI can add value to their operations. By setting clear objectives, organizations can avoid the pitfalls of implementing AI for the sake of AI and instead focus on the value it can bring (Ammanath, 2022).

Developing a roadmap for AI integration is another crucial step toward becoming AI-ready. This includes laying out a phased approach for AI implementation, identifying the resources needed, and establishing cross-functional teams to drive AI initiatives. The roadmap should serve as a guide, detailing the steps the organization needs to take to integrate AI successfully and become a genuinely AI-ready organization. These themes are discussed in much detail in later chapters.

Creating a Supportive Culture and Fostering Collaboration

Successful AI implementations heavily depend on a supportive culture and collaboration. Such a culture fosters an environment of learning, experimentation, innovation, and trust among employees, encouraging acceptance of new technologies and change. Leaders are instrumental in cultivating this supportive atmosphere. Their commitment to learning, openness to novel ideas, and readiness for collaboration can significantly foster an organizational culture prepared to embrace AI (Bruhn, 2022).

Promoting a supportive culture extends beyond nurturing a learning mindset; it necessitates the elimination of silos and the promotion of cross-team collaboration. It is about creating an environment where learning is cherished, knowledge and ideas are shared openly, and teamwork becomes the rule, not the exception (English, 2023).

In AI, collaboration takes special prominence as it often demands the harmonious interaction of diverse teams, such as data scientists, IT professionals, and business leaders. The collaborative synergy between these different skill sets and perspectives is crucial to achieving a common goal (English, 2023).

AI also plays a vital role in fostering a culture of collaboration within organizations. With AI-powered systems such as virtual assistants or chatbots, teams can receive real-time feedback and suggestions for improvement. These tools automate routine tasks and facilitate knowledge-sharing across teams, irrespective of their physical location. They can bridge gaps between teams in different geographical locations, making collaboration more efficient.

AI systems not only support collaboration but can also facilitate it across geographical boundaries; they serve as highly efficient communication and collaboration tools facilitating virtual collaboration for teams spread out across different locations. These systems can manage ideas, and project timelines, and track progress, ensuring everyone stays synchronized, irrespective of geographical location.

By leveraging AI systems for real-time collaboration, organizations can guarantee a smooth exchange of ideas and knowledge, creating an environment of shared learning and continuous improvement. This enhances productivity, cultivates an inclusive culture of innovation, and ensures that every voice is heard and every idea is evaluated.

AI systems are not merely supportive tools; they are active contributors to the innovation process. Utilizing AI's unique capabilities in idea generation, concept refinement, and collaboration facilitation allows organizations to nurture a sturdy and inclusive innovation culture. This drives growth and a competitive edge in the digital era.

Cultivating Innovation: Driving Creativity with AI

AI has emerged as a catalyst in an evolving business world, fostering innovation, agility, and forward-thinking. Adopting tools and technologies that enable rapid ideation, intelligent problem-solving, and informed decision-making is critical (Haefner et al., 2020).

Redefining the traditional idea-generation process, AI augments human intelligence and expands what is possible. With technologies such as Natural Language Processing (NLP) and Machine Learning (ML), AI revolutionizes how ideas are generated, providing unique perspectives free from human biases. The transformative nature of this capability becomes evident in AI-assisted brainstorming sessions. Participants can observe how an AI tool provides real time suggestions, utilizing a vast pool of data to offer unique combinations of concepts that may not have naturally occurred to them. These AI solutions can analyze extensive text data, uncovering insights that spark inspiration.

AI's impact on ideation is not restricted to facilitating brainstorming but also extends to fostering a diverse and innovative culture within organizations. By incorporating AI into the idea-generation process, organizations boost their creative potential, increasing the odds of producing groundbreaking innovations. This inclusive approach enables innovation from any level within the organization (Haefner et al., 2020).

Furthermore, AI's potential to present a range of perspectives from its vast knowledge base can foster diversity of thought. This can lead to a broader array of ideas, minimizing the risk of groupthink, and promoting an inclusive ideation process that leverages the diverse knowledge and skills within an organization.

AI also brings systematicity to ideation, capturing, storing, rating, and sorting ideas based on specific criteria, ensuring no potential innovation is overlooked. This systematic approach enhances the efficiency and effectiveness of the ideation process.

AI systems have progressed from being simple tools to becoming essential partners in the process of innovation. They possess the ability to examine vast amounts of data, recognize patterns, and produce valuable insights. These capabilities can greatly enhance the generation of ideas, the polishing of concepts, and the optimization of solutions (Haefner et al., 2020).

One way to integrate AI into innovation is through AI-powered virtual assistants or chatbots. These digital counterparts can assist teams in various ways, such as participating in ideation sessions or suggesting improvements during the concept development stage, thereby enriching the ideation process.

AI assistants can also handle repetitive tasks such as data entry, reporting, or scheduling, allowing human team members to concentrate on strategic and creative tasks. By taking over these routine tasks, AI systems allow human cognitive resources to focus on innovation, fostering a more efficient and creative working environment.

Integrating AI into the Innovation Process

Organizations must effectively integrate AI into their innovation processes to stay ahead in a market that thrives on disruption. This incorporation involves a two-pronged approach. First, it necessitates the deployment of AI technologies. Second, it requires the alignment of AI initiatives with the broader goals of innovation. When correctly implemented, these two aspects can transform how organizations innovate (Haefner et al., 2020).

AI can facilitate rapid prototyping, an essential step in innovation. Traditional prototyping methods can be time-consuming and resource-intensive. AI-powered tools can help streamline this process by generating and evaluating numerous design alternatives in a fraction of the time it would take humans. By using AI in prototyping, organizations can accelerate their innovation cycles, test more ideas, and get their products to market faster (Haefner et al., 2020).

AI can also significantly enhance the testing and validation phase of the innovation process. This phase often involves manual work in traditional settings, like sorting through user feedback or analyzing product performance data. AI can automate these tasks and carry them out with incredible speed and accuracy. This allows organizations to validate their ideas quickly and efficiently, enabling them to iterate and refine their innovations more effectively.

Additionally, AI tools have the capacity to foster inclusivity by guaranteeing that the perspectives of each and every individual are duly acknowledged. They can gather and analyze ideas from all members of an organization, providing a diverse range of perspectives considered during the innovation process. This democratic approach to innovation can drive better outcomes and cultivate a culture where innovation is everyone's responsibility.

Integrating AI into the innovation process is a game-changer. AI tools hold immense potential in accelerating innovation, streamlining processes, and nurturing collaboration and inclusiveness within organizations, empowering organizations to stay at the forefront of their industries, fully prepared to confront forthcoming challenges.

Embracing an Agile Mindset

Embracing AI requires transitioning from traditional, linear approaches to a mindset that champions agility, continuous experimentation, and lifelong learning. Introducing AI into business operations is not an isolated event or a one-time project; it signifies the beginning of a continuous journey of exploration, adaptation, and growth (Lee et al., 2020).

An agile mindset fosters a culture of flexibility and adaptability, critical elements for thriving in an AI-enabled business environment. This mindset encourages continuous iteration and refinement, where ideas are tested, validated, and quickly improved. This iterative process, supported by AI's capability for rapid data processing and generating insights, allows organizations to respond swiftly to market changes, thereby maintaining a competitive edge (Lee et al., 2020).

Experimentation lies at the heart of an agile mindset. With their predictive modeling and simulation capabilities, AI technologies provide a safe environment for such experiments. Companies can use AI to model different scenarios, predict outcomes, and assess the potential impact of their innovations without significant upfront investment or risk. This environment encourages teams to try unconventional ideas, learn from their attempts, and iterate, ultimately leading to breakthrough innovations.

Continuous learning is another cornerstone of an agile mindset in the AI era (Neeley & Leonardi, 2022). AI continually evolves, with new models, techniques, and applications emerging regularly. Organizations must commit to ongoing learning and development to stay abreast of these changes and leverage AI's full potential. This commitment might involve providing employees with training in AI and related technologies, staying updated on the latest AI research and developments, and learning from the AI implementation experiences of other organizations in their industry.

Embracing an agile mindset also involves learning from both successes and failures. Whether or not every experiment yields the desired result, it offers valuable insights that can inform future strategies and decisions. AI can aid learning by providing detailed analytics and insights about each experiment's outcomes, helping organizations understand what worked, what did not, and why (Lee et al., 2020).

Adopting an agile mindset in the AI era positions organizations for sustainable innovation and growth. By welcoming agility, promoting a culture of continuous experimentation and learning, and leveraging the potent capabilities of AI, businesses can navigate the complexities of the digital age, seize opportunities, and drive transformative innovation.

Cultural Transformation

Implementing AI within an organization often necessitates a significant cultural shift. This shift involves transitioning to a mindset that embraces data-driven decision-making, continuous learning, and innovation. As the custodians of organizational culture, business leaders play a pivotal role in guiding this transformation (English, 2023).

Promoting Data-Driven Decision-Making: A successful AI implementation requires the organization to adopt data-driven decision-making. Business leaders can foster this change by consistently utilizing data in their decisions and demonstrating the benefits of this approach. They can also promote organizational data literacy, enabling employees to understand and use data effectively.

Fostering a Culture of Learning and Adaptation: Continuous learning is crucial with the rapid pace of AI development. Business leaders need to encourage ongoing learning and skill development. Leaders can support their teams by providing access to relevant resources, bringing in external experts for knowledge-sharing sessions, and supporting training initiatives. Leaders must model adaptability by showing that changing course based on new information or learning is acceptable (Neeley & Leonardi, 2022).

Recognizing and Rewarding Effort: Recognition and reward systems should be revised to reflect the new skills and roles involved in AI initiatives. Celebrating successes, acknowledging the hard work, and incentivizing employees to participate in AI projects can boost morale and encourage further involvement.

The cultural transformation required for successful AI implementation is a strategic journey that business leaders must guide in an environment conducive to its adoption as part of the DNA fabric of the organization.

Driving Transformative Change with AI

AI's impact on innovation goes beyond individual processes or products. It has the potential to drive transformative change within organizations and across industries. By integrating AI into their innovation processes, organizations can more effectively address customer needs, navigate the rapidly evolving business world, and maintain a competitive edge (Ammanath, 2022).

AI-powered innovation can unlock new growth opportunities and position organizations as industry leaders. By understanding and leveraging the unique capabilities of AI, businesses can catalyze innovation and shape the future of their industries.

Conclusion

Organizations prioritizing building an AI-ready culture will be better equipped to adapt, flourish, and shape the future of business. By following these principles, organizations can lay a strong foundation for AI integration, empowering them to make data-driven decisions, drive innovation, and stay competitive as AI continues to revolutionize industries and redefine work.

Chapter 4

BUILDING YOUR AI STRATEGY AND TEAM

In today's rapidly competitive environment, the strategic consideration of AI is a business imperative. Incorporating AI into business operations goes beyond just adopting cutting-edge technology. It requires a profound understanding of AI's potential impact and the business's readiness to embrace a transformative change. Successful AI implementation will ultimately depend on identifying where AI can add value and aligning this with the organization's strategic goals (Mohan, 2022).

For instance, a company overwhelmed with routine customer inquiries may greatly benefit from an AI-powered chatbot, while another experiencing inventory management difficulties could leverage AI for predictive analytics. Thus, recognizing the areas where AI can provide substantial value is vital for successfully implementing and maximizing its potential advantages.

Businesses should start by thoroughly analyzing their operations, products, and services. The objective is to pinpoint opportunities where AI can boost efficiency, increase customer satisfaction, spur innovation, or even drive new opportunities. Whether automating mundane tasks, processing massive data volumes, personalizing customer interactions, or predicting future market patterns, AI offers a range of potential applications.

Afterward, the focus should shift toward evaluating the practicality and feasibility of AI integration into these identified areas. Data availability, resource and skill access, potential risks, and effective risk-mitigation strategies demand careful consideration.

Having completed this initial process, businesses should have a clear roadmap for launching their AI journey. The following steps involve carving out an AI strategy, which includes setting specific AI goals, crafting an implementation plan, defining steps for achieving these goals, and establishing metrics for measuring progress. Moreover, businesses must ensure sufficient financial support for the entire AI program lifecycle, from inception, and development to deployment, maintenance, support, and future enhancements.

Developing an AI Strategy

An efficient AI strategy is imperative to accomplish the objectives of an organization, while aligning with its values and culture. To construct a strong AI strategy, the following critical elements need consideration to align with the overarching business strategy:

Objectives: It is imperative, first, to define the desired outcomes. These might be boosting sales, elevating customer satisfaction, and cutting costs to driving innovation. Setting clear goals for AI integration into business operations is necessary.

Resources: Subsequently, and equally important for a successful AI strategy, the required resources need to be identified. This encompasses data, technology, specialized expertise, and financial allocations. A comprehensive understanding of the required resources is crucial to ensure the seamless execution of AI initiatives.

Gaps and Opportunities: The organization's current capabilities will need to be assessed and compared against gaps where AI can address critical business challenges or unlock new opportunities. This requires collaboration with domain experts, business leaders, and data science teams to identify areas where AI can have the most significant impact. In addition, the business needs to prepare and explore use cases, conduct pilot projects, and evaluate AI applications' feasibility and potential value in your industry and specific business context.

Risks and Challenges: Maintaining a keen awareness of potential obstacles and risks of AI adoption is essential. Recognizing these challenges in advance enables the development of strategies to mitigate them effectively.

Metrics: Defining metrics to measure the success of AI implementations should be done early on. These could be key business performance indicators (KPIs) such as revenue growth or customer satisfaction or AI-specific metrics like algorithm accuracy or processing speed. The selection of suitable metrics enables the monitoring of progress and the implementation of necessary adjustments (D'Silva & Lawler, 2022).

Establish Collaborative Approaches: Fostering collaboration and close alignment between business teams, IT, and data science teams is critical to build a cohesive team. Creating a cross-functional team to ensure a shared understanding of AI goals and implementation strategies could be the difference between success and failure. This requires open communication and knowledge sharing to harness the expertise of different stakeholders and maximize the effectiveness of AI initiatives.

Continuous Monitoring and Evaluation: Lastly, establishing mechanisms for monitoring and re-evaluating the success of AI initiatives will be the enabler of a virtuous cycle of continuous improvements. Regularly tracking and measuring KPIs to assess the impact of AI on your business objectives should become part of the apparatus being built. Collecting feedback from end-users and stakeholders to identify areas for improvement and refinement will only improve the outcomes and eventually help optimize your AI strategy to ensure alignment with evolving business needs (D'Silva & Lawler, 2022).

Aligning AI initiatives with the business roadmap ensures that AI is not implemented in a vacuum but is purposefully integrated to address specific business challenges, deliver value, and create a competitive advantage. Regularly revisit your business roadmap and adjust your AI strategy to ensure ongoing alignment with your organization's evolving goals and market dynamics (Vartak, 2022).

Prioritizing AI Initiatives

Business leaders must consider multiple factors when prioritizing which AI initiatives to tackle first. First, aligning AI projects with the overarching business goals is paramount, ensuring they catalyze the organization's strategic objectives (Gurtu, 2021).

Second, the leaders should assess the feasibility of a project. Some projects require a substantial allocation of infrastructure, expertise, and time, while others might provide short-term returns with minimal investments. Striking a balance in the portfolio of AI projects becomes essential. Project leaders need to dedicate resources to high-priority projects while maintaining a balance of short-term and long-term initiatives.

Third, it is crucial to consider the possible return on investment (ROI). While quantifying the ROI of AI projects may pose challenges, a ballpark figure of potential cost reduction, revenue boosts, and efficiency improvements can offer valuable perspectives for decision-making (D'Silva & Lawler, 2022).

Finally, risks tied to the AI project demand careful consideration. These include technical risks (such as technology not delivering as predicted), operational risks (like the potential disruption of business operations during rollout), and strategic risks (like the project failing to provide expected strategic advantages) (Ng, 2019).

By considering these factors, business leaders can efficiently prioritize their AI initiatives and ensure their AI investments yield the highest possible value.

Building an AI Team

The implementation of AI requires a specialized skill set and knowledge base. Depending on the needs and resources of the organization, an internal AI team could be established, external consultants hired, or AI service providers utilized (Ng, 2019). Regardless of the chosen approach, several crucial roles might be needed in the AI division, contingent on the project's scale:

Data Scientists: These professionals are crucial as they build and refine AI models. A solid foundation in machine learning, mathematics, statistics, and programming is vital.

Data Engineers: These individuals manage the data architecture, flows, and integrations. Mastery of databases, data lakes, data management skills, and data security is essential.

AI Ethicist: This role ensures that AI operations are conducted ethically and responsibly. A background in AI, ethical practices, compliance, and law is necessary.

The individuals in these roles should closely collaborate with business stakeholders, who will be the end-users of AI solutions, and make decisions based on the insights provided by AI (Nguyen, 2022).

Other roles, such as software engineers, quality assurance engineers, user interface and experience designers, and business analysts, are vital for ensuring the final product is functional and stable.

Selecting Suitable AI Technologies

AI projects often start on a small scale, serving as proof-of-concept or addressing a specific business need. However, as the benefits of AI become more evident, there is usually a desire to expand these projects quickly and apply their learnings to other areas of the business or increase the scope of existing projects. Therefore, systems must be able to evolve and scale, not only in terms of their training model and performance but also by the level of support available from engineering and operational teams. Leaders need to ensure they develop frameworks, systems, and processes that are modular and adaptable so AI systems can grow and evolve with the business.

Many AI capabilities are available, each with their strengths and weaknesses. Some are readily available to use or integrate, while others require further customization and configuration. The selection should be based on the organization's needs and resources.

Modularity: AI systems should be designed or chosen for their modularity, allowing different components to be added, removed, or upgraded independently. This fundamental flexibility enables the system to adapt as the business matures, whether by amplifying its capabilities or integrating with other systems.

Adaptability: As business needs change and evolve, AI systems should be able to accommodate new data types and structures, adapt to new business processes, and integrate with other emerging technologies. Adaptability ensures the system stays pertinent over time without necessitating considerable additional investments or major changes.

Performance: As the scale of data and the complexity of tasks increases, the AI system must continue to perform effectively by having adequate processing power, data storage, and the ability to apply AI algorithms efficiently. If the AI service is external, close attention is needed for ongoing costs and maintaining these services. For example, if hosted in a cloud environment, cost management becomes critical to ensure that compute and storage spending aligns with set budgets.

Functionality: Does the technology serve the specific needs and capabilities of the organization? For example, the technology supporting natural language processing would be suitable if text data analysis is required.

Ease of use: Is the technology user-friendly? While some AI technologies demand extensive coding and technical expertise, others offer a more user-friendly approach, enabling integration via Application Programming Interfaces (APIs) or available as plug-ins.

Scalability: Is the technology capable of handling the existing volume of data and adaptable to future needs? Furthermore, some solutions might employ a usage limit due to rising demands. As of May 2023, for example, ChatGPT 4.0 restricts usage to 25 prompts every three hours, while the API version has limits on the input and output text length.

Cost: What is the expense associated with the technology? The initial investment and ongoing costs for maintenance and upgrades should be considered. These expenses might include a subscription fee, a per-seat cost, or the computational cost of running a machine-learning model in the cloud.

Building and Training AI Models

After identifying opportunities, developing an AI strategy, forming an AI team, and selecting suitable AI technology, the next phase involves constructing, training, or customizing AI models. Data scientists and engineers play a pivotal role in this process (Williams & Mattar, 2022).

Data Collection and Preparation: AI models are data-driven. This step encompasses gathering pertinent data and preparing it for use in AI models. Preparation might include data cleaning, managing missing attributes, augmenting data with supplementary context, and transforming the data into a format compatible with the model. For text completion or text analysis use cases, this step may be optional (Walch & Schmelzer, 2021).

Model Selection: An AI model fitting the specific task must be chosen. The choice depends on the nature of the data and the problem that needs solving.

Model Training: This stage involves using the prepared data to train the AI model. The model is input with this data, and its learning accuracy should be verified (Walch & Schmelzer, 2021).

Model Testing: Once the model is trained, it must be tested to evaluate its performance on new data. To measure the model's effectiveness, a separate data set (not used in training) is employed for testing purposes.

Model Deployment: Upon satisfactory performance of the model, it can be deployed to generate predictions or make decisions based on new data.

Executing a Pilot Test

With an AI strategy developed, implementing a methodical approach is prudent. Before initiating full-scale AI implementation, it is sensible to commence with a pilot test or a prototype. This preliminary step allows for a calculated assessment of the AI solution's compatibility with the current business framework, culture, and infrastructure, while also unveiling potential challenges (Ng, 2019).

The pilot test should be narrow in scope, targeting a particular business issue, and utilizing a manageable volume of data. This strategy allows close monitoring of the AI solution's impact and the ability to make necessary adjustments without causing large-scale disruption.

A pilot test is not merely a trial run. It is an iterative process that allows the AI model to learn, improve, and fine-tune its accuracy with each cycle. This capacity for continuous improvement is at the heart of AI and machine learning. Executing a pilot provides the opportunity to explore the tangible implications of AI - how the technology will interact with users, adapt to evolving business conditions, and achieve the intended objectives. More importantly, it allows the business to tweak its approach before large-scale investment (Walch & Schmelzer, 2021).

Ensuring all relevant stakeholders understand the pilot's objective and the evaluation metrics is critical to a fair appraisal of the AI performance. Essential elements to consider for a successful pilot include stakeholder buy-in, data availability, technology compatibility, and the organization's preparedness in terms of relevant skills and adaptability.

After concluding the pilot test, take the time to review the results, seek feedback, and identify successes and failures. This is also an opportune time to gauge AI's potential impact on customers, employees, and business procedures and refine your broader AI strategy accordingly. A pilot test is an effective risk-management tool, setting the stage for successful and value-enhancing AI implementation.

Deploying AI Solutions

Having proven the feasibility of AI solutions with a pilot and ensured the right investments are in place, the next step is to deploy these solutions within the organization. This is a multifaceted process requiring diligent planning, communication, change management, and execution.

Integration: To integrate AI solutions into existing systems and procedures, adjusting these systems to accept input from AI models and react based on the results may be necessary. This requires a close alignment with product management, IT staff, or third-party providers to allow ample time for a successful implementation.

Testing: Rigorous testing of AI solutions before launch is vital to guarantee optimal performance. While some testing might have taken place during a pilot phase, testing in a production-like environment, often called "staging," ensures proper production readiness.

Training and Support: Providing ample training and support to users is essential. This empowers them to fully utilize and comprehend the AI tools, thereby maximizing productivity.

Monitoring and Improving AI Performance

The application of AI is not a one-time occurrence but an ongoing process that requires regular monitoring and improvement to ensure that AI solutions continue to deliver value in line with shifting business priorities and mandates (Walch & Schmelzer, 2021).

Monitoring: Continuous tracking of the AI solutions' performance is necessary. Keeping an eye on important metrics, scrutinizing user feedback, and conducting routine checks are critical for success.

Maintenance: Regular upkeep ensures the ongoing functioning of AI systems. To enhance user experience, actions such as fine-tuning AI models, rectifying issues, and improving data accuracy may be undertaken.

Improvement: Continuous advancement in AI solutions necessitates a proactive stance. It is vital to refine AI models continuously, expand the range of projects, and incorporate cutting-edge AI technologies. This strategy aids in maintaining a competitive edge and achieving tremendous success.

Reevaluation: The AI strategy should be periodically reassessed, considering new developments in AI technology and changes in the business landscape.

Role of Business Leaders in AI Initiatives

Business leaders play a crucial role in successful AI implementation. Their responsibilities go beyond making financial investments and involve strategic and operational leadership, cultural transformation, and ethical governance.

Strategic Leadership

Aligning AI initiatives with the business's overarching strategy lies primarily with business leaders. They need a solid understanding of AI technologies' potential and limitations to make informed decisions about where best to allocate resources (Fountaine et al., 2020).

Business leaders must identify AI opportunities aligning with business objectives and offering strategic advantages. To achieve this, they must stay updated on AI advancements, trends, and their competitors' activities.

Effective strategic leadership also includes prioritizing AI projects based on their potential impact, feasibility, and alignment with the business's strategic goals. Leaders may need to make tough decisions to focus on a few high-priority projects rather than spreading resources too thinly across many initiatives.

Business leaders also play a crucial role in allocating the necessary resources effectively. This involves not only financial investment but also the right human capital - skilled data scientists, engineers, and project managers who can translate AI potential into reality. They also need to establish clear accountability structures for AI project execution. They do this by setting measurable goals, defining roles and responsibilities, and installing a system of checks and balances. This involves setting up clear reporting lines and performance metrics, ensuring that everyone involved in the AI project knows what they are accountable for, and that progress is continually assessed and realigned with the overall business objectives (Fountaine et al., 2020).

Operational Leadership

Operational leadership is pivotal in the successful execution of AI projects. This role calls for a blend of project management, resource optimization, risk mitigation, and performance tracking, all while maintaining an agile and responsive approach to the dynamics of AI implementation.

Cross-Functional Coordination: AI initiatives typically involve diverse stakeholders, including data scientists, IT professionals, business analysts, and end-users. As such, business leaders must establish clear communication channels and foster a collaborative environment to ensure cross-functional alignment. One way to ensure effective teamwork is by clearly defining each team member's roles and responsibilities, promoting mutual understanding and respect, and resolving conflicting priorities or interests.

Resource Management: AI projects require careful management of resources, including data, technology, and human talent. Business leaders must ensure that the necessary resources are available when and where they are needed and used efficiently. To successfully manage AI projects, tasks such as securing data access, optimizing computational resources, and distributing the workload among AI specialists must be negotiated and prioritized.

Risk Management: AI initiatives can carry significant risks, such as technical failure, data breaches, or unanticipated consequences of AI decisions. Business leaders need to identify these risks upfront, put measures in place to mitigate them, and have contingency plans ready for when things go wrong. A proactive approach to risk management can help avoid costly and disruptive setbacks.

Performance Tracking and Adjustment: With AI projects' complex and experimental nature, ongoing performance tracking is essential. Business leaders need to establish key performance indicators (KPIs) that align with the goals of the AI initiative and regularly review progress against these KPIs. Where performance falls short, leaders must identify the causes and make necessary adjustments - whether revising the project plan, providing additional resources, or tweaking the AI models.

Operational leadership, thus, requires a proactive, hands-on approach to guiding the AI initiative from inception to completion. Through effective coordination, resource management, risk mitigation, and performance monitoring, business leaders can ensure the smooth execution of AI projects and maximize their return on investment.

Overcoming Barriers to AI Adoption

Although the advantages of AI are apparent, businesses often need help attempting to implement this technology. These hurdles may involve more understanding of AI and its workings, requiring more skills and talent, worrying about data privacy and security, and difficulty incorporating AI with established systems and procedures (Goasduff, 2019).

Education and awareness: To counteract misconceptions and apprehensions about AI, businesses need to enlighten their employees about what AI signifies, how it operates, and how it can contribute value. This involves training and resources to assist employees in comprehending and employing AI tools.

Talent recruitment and growth: Businesses must attract and keep talent possessing AI and data science capabilities. This could mean hiring new employees, re-skilling existing ones, or collaborating with external providers (Walden, 2022).

Data management: Businesses must have the appropriate data infrastructure to facilitate AI. This entails investing in data storage and processing capacities, executing data governance policies, and ensuring adherence to data privacy laws.

Integration with current systems: Businesses must amalgamate AI with existing systems and procedures. This demands a clear implementation strategy and sustained support to guarantee that AI tools are effectively utilized.

Cultivating Transparency and Trust: Implementing AI systems can lead to apprehension and resistance among employees, especially if there are concerns about job displacement or privacy. Leaders can alleviate these fears by maintaining transparency about the organization's AI strategies, the potential impact on jobs, and measures to ensure data security. Cultivating trust is crucial for accepting and adopting AI within the organization.

Long-Term Considerations for AI Adoption

While beginning AI integration within a business, leaders must look beyond immediate needs and challenges. As AI takes hold, leaders need to ensure they have a framework for a virtuous cycle of growth and scalability and continuous learning and advancement.

Continuous Learning and Adaptation

AI is a dynamic field marked by continuous evolution and improvement. Hence, integrating AI into business operations is not a one-and-done process but an ongoing journey that requires constant learning and adaptation.

Continuous Learning: AI models learn from data; their accuracy and reliability improve as they are exposed to more varied information. However, this is not a static process. Over time, as market conditions shift, customer behaviors evolve, and business needs change, the data patterns that AI models were initially trained on may no longer be relevant, which necessitates continuous learning, a process by which AI models are periodically retrained on new data, keeping them updated and accurate.

Model Monitoring: Alongside continuous learning, businesses must constantly monitor their AI models' performance by tracking key metrics such as accuracy, precision, recall, or other company-specific KPIs. Regular monitoring can detect performance degradation and identify when the model needs retraining.

Adaptive AI: Some advanced AI systems are capable of adaptive learning, where they can self-update in response to new data. While this capability can significantly enhance their ability to stay relevant, as they can automatically adjust to changing data patterns without human intervention, these systems still need oversight to ensure they adapt correctly.

Learning Culture: Lastly, this principle of continuous learning and adaptation also extends to the human aspects of a business. Employees need to be upskilled to work with AI systems effectively. The rapid pace of AI advancements means continual education and training should be part of the business culture (Walden, 2022).

Businesses must invest in an ongoing learning, monitoring, and adaptation process to ensure their AI systems stay relevant and practical. AI can remain a valuable asset supporting the business's growth and evolution with this approach.

Social Responsibility

Integrating AI into business operations presents significant opportunities and carries essential social responsibilities. The widespread use of AI has societal implications that businesses must consider as they design and implement AI systems. These include potential job displacement due to automation, risks of bias in AI algorithms, and ethical use of data.

As businesses increasingly adopt AI, they must consider not just the economic implications but also the social ones. By addressing these issues proactively, they can build trust with their customers, employees, and other stakeholders and position themselves as responsible and ethical users of AI technology.

Chapter 5

RESPONSIBLE AI: ETHICS, PRIVACY, AND SECURITY

Balancing Technological Advancements and Ethical Responsibilities in AI's Adoption

Artificial Intelligence (AI) has emerged as a transformative force across industries, revolutionizing how businesses operate and interact with customers (Fugulin, 2023). At the heart of AI's capabilities lies data – vast amounts of information that fuel machine learning algorithms, enabling AI systems to learn, make decisions, and deliver value. However, with great power comes the great responsibility of safeguarding data, ensuring its privacy and security, all while ensuring the implications for individuals, businesses, and society are handled ethically.

Protecting Individual Privacy in AI

In the digital age, data has become the primary currency of exchange. Data does not merely reflect personal attributes; it provides intimate details of users' tastes, preferences, habits, and behaviors. This data offers a goldmine of insights for businesses, enabling them to understand their customers better and tailor their services to match individual needs. AI systems, with their powerful data processing capabilities, facilitate this understanding. However, this also emphasizes the need to safeguard personal privacy ("Putting Principles into Practice," 2021).

Respecting privacy in AI's implementation is not just about meeting legal requirements or risk mitigation—it is about earning and retaining customer trust. In an increasingly digital world, where data breaches are a persistent threat, privacy assurance can serve as a differentiator, setting businesses apart from their competitors. Organizations that put privacy at the heart of their AI endeavors are committed to their customer's interests. By doing so, they cultivate trust, fostering stronger relationships and enhancing customer loyalty.

To operationalize this commitment to privacy, businesses must adhere to the principles of "privacy by design," embedding privacy protections into their AI systems from the outset. This involves ensuring that AI models are designed to process minimal personal data and that any processed data is anonymized. Moreover, transparency is critical. Businesses should communicate to customers how AI systems use and process their data, giving customers control over it.

Mitigating Risks of Data Breaches and Cyberattacks

The massive amounts of data can become enticing targets for cybercriminals. Data breaches can lead to substantial financial losses and cause significant reputational damage, eroding the trust businesses have worked hard to build. As such, companies must treat cybersecurity not as an afterthought but as an integral component of their AI systems.

This involves investing in robust cybersecurity measures, including solid data encryption, secure storage solutions, stringent access controls, and advanced threat detection systems. These measures can provide a particular line of defense, protecting AI systems from potential cyberattacks. Cybersecurity cannot be a one-off effort; instead, it must involve ongoing vigilance, regular security audits, and proactive monitoring of emerging threats.

Furthermore, employees play a vital role in maintaining cybersecurity. By training employees on data protection best practices, businesses can reduce the risk of inadvertent breaches. Companies can foster a culture of cybersecurity, embedding cybersecurity principles into their organizational DNA. This includes ensuring accountability at all levels and incentivizing employees to prioritize cybersecurity.

Strategies for Ensuring Data Privacy and Security

As we navigate the AI era, protecting sensitive data and ensuring robust security measures are paramount (Yaqoob, 2023). Businesses are grappling with the increasing amount of data generated, and understanding how to handle this data responsibly and securely is critical.

To prioritize data sensitivity, the first step is to identify and classify the data that businesses handle. This process entails understanding what data is sensitive, who has access to it, where it is stored, and how it is shared. Such understanding allows businesses to construct tailored strategies to secure their data. The next step in this process is securing the data. Techniques like robust encryption, involving strong algorithms and secure critical management systems, can "ensure that even if unauthorized individuals gain access to the data," they cannot read or misuse it. Regular data backups are another essential strategy. They should be conducted both on-premises and off-site, mitigating the risk of data loss due to hardware failures, natural disasters, or cyber incidents (May, 2023).

Regulatory Compliance in Data Privacy and Security

Beyond these technical measures, regulatory compliance is pivotal in data privacy and security. As data privacy laws and regulations grow increasingly complex, businesses must actively seek to understand and adhere to these laws. Non-compliance can result in substantial fines, legal repercussions, and severe damage to the business's reputation. Organizations must continuously monitor the evolving regulatory environment, take proactive measures to meet these requirements, and build robust frameworks for reporting data breaches.

To navigate these intricate pathways of regulatory compliance effectively, businesses must cultivate a strong relationship between their technology and legal teams. Regular collaboration between these teams is essential. This ensures that technology strategies and business models align with privacy laws and security regulations, leading to a culture of compliance.

Establishing a Culture of Data Privacy and Security

Building a data privacy and security culture goes beyond adopting technology solutions and complying with regulations. It is about creating an environment where every organization member understands the importance of data privacy and security and is responsible for protecting sensitive information.

At the heart of this culture is leadership commitment. Leaders must set the tone by prioritizing data protection and demonstrating their commitment through action. This includes ensuring adequate resources for privacy and security initiatives and integrating these priorities into the business's strategy and goals. When leaders champion data privacy and security, it sends a powerful message to the entire organization about the value placed on these aspects.

Clear and comprehensive policies and procedures form the backbone of a robust privacy and security culture. These policies should provide clear guidance on data handling, access controls, incident response, and regulation compliance. They must be regularly reviewed and updated to reflect the evolving technological and regulatory frameworks, ensuring their relevance and effectiveness.

However, more than policies are needed; ongoing education and awareness programs must complement them. These programs aim to build a deep understanding of data privacy and security among all employees. Topics may include data classification, secure data handling, password hygiene, recognizing and avoiding phishing attempts, and responsible use of technology resources.

An effective incident response framework is critical in today's cyber environment. No defense is foolproof, and breaches can occur despite the best precautions. A solid incident response plan includes clear roles and responsibilities, communication protocols, and steps to mitigate the impact of a breach. It can significantly limit the damage from a breach and help the organization recover swiftly.

Finally, a robust privacy and security culture requires continuous monitoring and improvement. With the changing characteristics of technology and threats, there are other options than staying static. Regular audits, risk assessments, and feedback mechanisms can help identify gaps and improvement areas, keeping the organization's privacy and security posture aligned with evolving trends. The journey to a robust privacy and security culture is ongoing, but with commitment, collaboration, and continuous learning, it is worth taking.

Balancing Data Needs with Privacy Concerns

AI systems often consume vast amounts of data, some of which could be sensitive or personal. This inherent reliance on copious amounts of data places AI in a potentially precarious position vis-a-vis privacy rights, thereby giving rise to complex ethical dilemmas. This raises critical questions: Is it possible to balance the immense data demands of AI systems with the fundamental right to privacy? How can we ensure the sanctity of user privacy while meeting the data requirements of these systems (Yaqoob, 2023)?

In today's data-centric environment, these questions are relevant and pressing. They demand thoughtful reflection and the formulation of solid, effective strategies. While it is true that AI has the potential to infringe on privacy, the development of innovative privacy-enhancing technologies (PETs) indicates a positive trend toward reconciling this tension. These technologies aim to enable businesses to derive valuable insights from data without compromising user privacy. Techniques such as differential privacy and federated learning exemplify this approach.

Differential privacy is a method that adds noise to the data or outputs of an AI system, making it difficult to identify individual data points, thereby preserving user privacy. On the other hand, federated learning allows AI models to learn from data distributed across multiple devices or servers without moving or centralizing the data. These techniques can help businesses balance their data needs with privacy concerns, thus paving the way for ethical and responsible AI use.

This balance is not solely a technical challenge; it requires a philosophical and moral commitment to privacy and the recognition of its importance in our society. The AI community, including developers, businesses, and regulators, must adopt a holistic approach to tackling this issue, focusing on technical solutions and ethical guidelines, regulatory measures, and public awareness (Fugulin, 2023).

AI and Its Ethical Considerations

AI has evolved significantly since its inception in the mid-1950s. This progress was facilitated by two key factors: the availability of enormous computing power and the increasing ability to collect and analyze vast amounts of data.

Modern AI systems, often called "narrow" or "weak" AI, primarily use an inductive approach. They analyze large datasets, create or adapt decision rules, and optimize based on predefined criteria. These AI systems are adept at detecting and categorizing patterns and can adjust to new ways while in operation. However, their abilities are narrow and confined to the tasks they have been specifically trained to perform. This means an AI trained for one task can only be applied to a different task with further training and adaptation. This characteristic of AI leads us to one of the first ethical considerations: the challenge of algorithmic bias (Yaqoob, 2023).

Addressing Algorithmic Bias and Fairness

AI systems learn from historical data, which can carry biases, reflecting societal inequalities and prejudices. Consequently, AI systems can inadvertently perpetuate these biases, leading to unfair or discriminatory outcomes (Yaqoob, 2023). For example, AI systems employed in the hiring process have been shown to favor specific demographic profiles over others, propagating societal biases. Similarly, AI-driven loan approval systems have demonstrated discriminatory tendencies, biasing against certain racial or socioeconomic groups.

To mitigate bias and foster fairness, it is critical to adopt a multi-pronged approach. Firstly, it starts with the data. Businesses need to ensure that the training data used for AI systems are diversified and representative of the broad user base. This reduces the risk of any single group being disproportionately represented or neglected. Bias can also be mitigated through techniques like balanced weighting, where less-defined groups are given more weight during the training phase.

Secondly, businesses should conduct routine bias assessments, using fairness metrics to monitor their AI systems. These assessments can highlight potential biases, enabling firms to take corrective action. It is important to note that fairness is not a one-size-fits-all concept; different AI applications may require various fairness measures (Yaqoob, 2023). As such, businesses need to consider the context in which their AI systems operate, determining the most appropriate fairness measures.

Thirdly, businesses should foster transparency and interpretability in their AI systems. By making their AI models explainable, companies can better understand and communicate how these models make decisions (Yaqoob, 2023). This helps uncover hidden biases and fosters trust, as customers appreciate clarity and openness.

Data Privacy and Security

With the proliferation of AI in various sectors, an enormous amount of data is being collected, stored, and analyzed (Yaqoob, 2023). This raises significant ethical issues surrounding data privacy and security. Personal data is often used to train AI models, and without proper safeguards, there is a risk of misuse or breach of this sensitive information. Moreover, the extent to which personal data is used for AI training is often unclear to individuals, raising concerns about informed consent.

AI's ability to analyze and predict patterns in data can also be used in ways that infringe on individual privacy. For instance, predictive policing algorithms use data to forecast where crimes are likely to occur, which can lead to disproportionate policing of specific communities. Similarly, AI-powered surveillance systems can monitor individuals on a scale and to a degree of detail previously unimaginable, raising concerns about privacy rights and individual freedom.

Embracing Ethical AI Principles

AI can transform businesses and society, enabling efficiencies and innovation (Fugulin, 2023). However, this transformative power comes with ethical responsibilities. Companies must ensure that their AI systems align with societal values, principles, and norms. Embracing ethical AI principles is not just about risk mitigation—it is about acting as a responsible corporate citizen and contributing positively to society.

Businesses should engage with various stakeholders—employees, customers, regulators, and society—when developing their AI systems. This engagement can help to ensure that the AI systems are inclusive and reflect diverse perspectives.

Transparency and Explainability

Intense learning-based AI systems are often called "black boxes" due to their opaque inner workings. Their decision-making processes are complex, layered, and beyond human comprehension, fostering a need for more transparency. This opacity becomes a roadblock in scenarios demanding an understanding of why an AI system made a specific decision.

Transparency and explainability are critical in AI for multiple reasons. They foster trust between AI systems and their human users, which is crucial for the widespread adoption of AI technology (Fugulin, 2023). They are essential for diagnosing and rectifying errors in AI systems and crucial in sectors where decisions have serious consequences, such as healthcare or judicial systems.

However, the journey toward making AI systems interpretable is fraught with challenges. Complex models, like deep learning networks, inherently have thousands, if not millions, of parameters. Explaining their decisions in human-understandable terms is akin to untangling a formidable web of interconnections. However, numerous emerging approaches strive to shed light on the "black box." Techniques such as Local Interpretable Model-

Agnostic Explanations (LIME) or SHapley Additive exPlanations (SHAP) offer means to understand individual predictions, offering a peek into the reasoning of AI (Yaqoob, 2023).

Ethical Governance

As AI's use becomes more prevalent in business, ethical and governance issues become increasingly important to address proactively. Business leaders must ensure that the ethical use of AI within their organizations is a core foundation of acceptable use policy, training, and change management efforts. To ensure the responsible use of AI, it is also necessary to establish governance structures, implement ethical guidelines, and guarantee transparency and accountability in decision-making processes (Yaqoob, 2023).

By playing these roles effectively, business leaders can ensure successful AI implementation and unlock the full potential of AI for their businesses.

Accountability

Navigating the complex world of AI can make assigning blame for errors feel like searching for a needle in a haystack. Ensuring the responsible use of AI systems and appropriately addressing any harm they cause underscores the complex issue of accountability in AI—the questions surrounding responsibility reach the heart of our legal and regulatory systems (Banta, 2023).

Attributing responsibility for decisions made by AI systems is a complicated task. Multiple entities, including the developers of the AI, the users, and potentially even the AI system itself, may share responsibility ("Putting Principles into Practice," 2021). Our existing legal and regulatory frameworks may need a significant overhaul to meet these challenges.

There are several ways to ensure accountability in AI use, such as implementing legal and regulatory guidelines that outline clear responsibilities and obligations for all parties involved. Techniques like AI audits also help ensure accountability by assessing AI systems' safety, fairness, and robustness (Yaqoob, 2023). Ultimately, the path to accountable AI is a combination of technical, legal, and ethical efforts, continually evolving to address new challenges that may arise.

As AI systems become increasingly autonomous, determining accountability for their actions becomes more complex (Microsoft AI, 2021). This is particularly true for systems like autonomous vehicles, where decisions made by AI can have life-or-death consequences. Suppose an autonomous vehicle is involved in an accident, determining who is at fault. In that case, the vehicle's owner, the manufacturer, or the AI - is a complex question with significant legal and ethical implications.

Moreover, as AI systems make decisions traditionally made by humans - in fields as diverse as healthcare, finance, and criminal justice - ensuring that these decisions are transparent, fair, and accountable becomes increasingly essential (Yaqoob, 2023). However, the "black box" nature of many AI systems, where the decision-making process is opaque and difficult to understand, challenges this transparency.

Understanding its ethical implications is crucial as AI becomes more ingrained. From algorithmic bias to privacy concerns to accountability, these issues must be carefully considered and addressed to ensure that the development and use of AI align with our societal values and norms ("Building AI Ethics," 2023).

Ensuring AI Creative Writing Stays Within Ethical Boundaries

AI has also made its way into the creative world, with AI algorithms now capable of writing compelling content. However, when utilizing AI in creative writing, it is essential to ensure that it does not violate ethical standards, including plagiarism.

AI algorithms learn from the data they are trained on, which often includes vast amounts of existing content. Therefore, businesses must ensure that the AI's output replicates the training data sparingly and does not violate copyright laws or plagiarize others' work.

This can be achieved by implementing measures to detect and avoid plagiarism in AI's output, using techniques like text-matching software such as Copyleaks, ContentDetector.ai, or CrossPlag. Additionally, training the AI on diverse sources can reduce the likelihood of the AI replicating specific sources too closely.

Moreover, proper citation practices must be applied, where AI-generated content is based on or inspired by existing content. Acknowledging the sources of data AI was trained on, when feasible, could also contribute to transparency and ethical usage.

While AI holds significant potential for creative writing, its ethical usage is paramount. Businesses must implement necessary measures to ensure that AI respects intellectual property rights and maintains high ethical standards.

Conclusion

As the transformative power of AI accelerates, it is incumbent upon businesses to ensure that this technology is used responsibly. Balancing innovation with ethical responsibility, business leaders must aim to guide AI as a force for good, embodying societal values and principles. High data privacy and security standards, robust governance structures, and clear accountability are key to creating AI implementations that respect individual rights, promote fairness, and safeguard our collective future. Meanwhile, businesses must strive for transparency and explainability in their AI systems, fostering trust and making AI a technology that serves everyone, not just corporate interests (Yaqoob, 2023).

Furthermore, business leaders must recognize that privacy, security, fairness, and ethics are intertwined issues that mutually influence one another. A holistic approach to these concerns and a commitment to continually improve practices in line with evolving trends can help navigate the complex AI landscape.

Developing and using AI technologies calls for an ongoing ethical dialogue, a task that evolves alongside the technology itself (Fugulin, 2023). Strategies that echo societal values, uphold fairness, ensure justice, and minimize bias are crucial in this endeavor. Moreover, in the creative realms, upholding intellectual property rights and reinforcing ethical standards is essential.

Creating AI technologies with an ethical foundation ensures the development of sophisticated, equitable, and accountable systems. Viewing AI as a moral responsibility rather than merely a technological pursuit guarantees that AI serves as a tool enhancing human capabilities while abiding by ethical guidelines. This commitment to ethics paves the way for AI to empower individuals and society while adhering to fairness, transparency, and respect (Yaqoob, 2023). Simple, clear language devoid of exaggerated expressions aids this vital communication, driving the understanding and adoption of ethical AI among business and technology leaders.

Chapter 6

EXECUTING YOUR AI STRATEGY

The advent and rapid advancement of Artificial Intelligence (AI) offer vast opportunities for businesses across all sectors, driving unprecedented transformations in how organizations operate and deliver value to their customers. However, adopting and implementing AI effectively requires a holistic, strategic approach. Businesses must tackle a multitude of considerations, including but not limited to defining their AI strategy, building the necessary technical infrastructure, ensuring data privacy and security, maintaining regulatory compliance, fostering strategic partnerships, encouraging innovation, and embedding AI into the organizational fabric (Bergeret & Malaurent, 2023).

It is imperative that organizations comprehend the multifaceted nature of AI implementation. Businesses can capitalize on their AI investments by focusing on critical components such as setting up robust governance structures, adhering to ethical guidelines, nurturing innovation, and forming strategic collaborations. Moreover, upholding regulatory compliance and ethical practices while navigating the complex world of AI is crucial. Lastly, executing an AI strategy could potentially require significant investments across technology, infrastructure, talent development and acquisition, data management, and governance.

Technology and Infrastructure

Committing to AI often requires a significant uplift in a company's tech infrastructure and operational maturity. This demands careful planning, strategic funding, and, frequently, a considerable financial investment. The steps towards a solid foundation for AI comprise several key domains:

Cloud Computing Platforms: AI demands immense computational capacity, and cloud computing offers a scalable and efficient solution. Cloud platforms provide the backbone to run AI models, accommodate vast volumes of data, and enable faster processing and real-time analytics. Providers like Amazon Web Services, Google Cloud, and Microsoft Azure offer robust cloud solutions for AI deployment.

Specialized third-party providers that integrate with these cloud environments continue to spring up, offering a variety of options to choose from. For example, the Technology Innovation Institute (TII) in Abu Dhabi has announced its open-source large language model (LLM) with support for 40 billion parameters (Al Bannai, 2023).

Data Storage and Processing Capabilities: AI thrives on data. Large-scale data storage solutions, both on-premises and cloud-based, are critical for AI implementation. Furthermore, AI demands robust data processing capabilities. Investments might be needed in high-performance processors, servers, network bandwidth, and technology to manage and process big data like Hadoop or Spark.

AI Software and Tools: Many AI software and tools are available, both open-source and commercial. These include platforms for building AI models, like TensorFlow and PyTorch, data analysis and visualization tools, and AI applications for specific business functions. The choice of tools will depend on the company's particular needs and the AI skills available in-house (Marr, 2022).

Cybersecurity Measures: As companies integrate AI into their operations, they must strengthen their cybersecurity posture. Protecting the AI systems and the data they handle is essential. Companies may discover a need for investments in encryption technologies, secure data storage solutions, threat detection and response systems, and employee cybersecurity training.

Building a robust technology infrastructure for AI is a complex undertaking. It requires a clear vision, strategic planning, and ongoing investment. However, the potential benefits - improved efficiency, more informed decision-making, and the creation of new business opportunities - can be substantial. It is a journey worth taking for companies ready to embrace the future.

Data

Data undeniably fuel AI where AI-based systems lean heavily on extensive data quantities for learning and decision-making, making data a crucial aspect of any AI investment strategy. Therefore, the program should concentrate on three central areas: data procurement, management, and governance (Collins, 2022).

A robust data infrastructure is an important requisite for businesses looking to integrate AI, requiring establishing and strengthening data management and analytics capabilities to ensure the effective functioning of AI algorithms. This requires collecting, storing, processing, and analyzing a massive volume of data and a firm commitment to upholding data integrity, privacy, security, and compliance with regulations such as GDPR and CCPA ("An Executive's Guide to AI," 2020).

Data Acquisition: Obtaining reliable, relevant, and current data is crucial for a successful AI project, whether it originates from internal or external sources. These sources, housing structured or unstructured data, could include ERPs, CRMs, intranets, customer interactions, third-party data providers, public datasets, and other relevant sources. Companies may need to invest in data collection tools (web scraping or harvesting tools) and services or partner with providers to obtain the necessary data (Sagiraju, 2022).

Data Management: Once data is ingested and aggregated, it needs to be cleansed, classified, and prepared for use by AI algorithms. Data management can be a complicated endeavor comprising multiple steps along the way. These steps include data correction, whether automated or manual, data standardization, augmentation with attributes, and transformation into a suitable format for system analysis. Businesses should use data management tools and platforms to streamline this process. Hiring professionals specializing in data management can help oversee the process and ensure its success (Davenport & Redman, 2022).

Data Governance: Effective data governance ensures that data is used and handled appropriately. When managing data, it is essential to establish policies and procedures that address data privacy, security, compliance, quality, and lifecycle management. Therefore, businesses must invest in data governance by creating policies and procedures, installing data governance software, and training employees on proper data handling practices (Bergeret & Malaurent, 2023).

Data Security and Privacy: Security becomes critical as data volume and complexity grow. Investments may be needed in encryption, secure data access, and privacy-preserving technologies. Companies should also consider data anonymization techniques to protect individual privacy in datasets (Davenport & Redman, 2022).

Investing in a comprehensive data strategy is imperative for successful AI implementation. It ensures that AI systems have high-quality data to function effectively and that data is handled responsibly and ethically. Ultimately, a solid data strategy lays the foundation for a company's AI initiatives and helps unlock the transformative potential of AI (Bergeret & Malaurent, 2023).

Talent

AI projects' success relies heavily on assembling a diverse team, encompassing data scientists, machine learning engineers, data engineers, and business analysts. As the demand for these skills is high, businesses face a significant challenge in talent acquisition (Cohen, 2023). Therefore, investments in recruitment, training, and retention initiatives are paramount. A culture encouraging collaboration between technical and non-technical team members is also crucial for successful AI implementation.

Talent Acquisition: Building a strong AI team demands proactive talent acquisition strategies. Businesses should consider collaborating with tech-focused recruitment agencies, attending job fairs, offering internships, or seeking out potential candidates directly. Competitive compensation packages and clear career progression paths are essential to attract top talent (Dawson, 2020).

Training and Upskilling: Skills for AI implementation need not be excluded from new hires. Existing employees can be trained and upskilled to develop AI competencies. This can be facilitated through workshops, online courses, mentorship programs, and collaborations with educational institutions.

Retention Initiatives: Given the high demand for AI skills, retaining talent can prove as challenging as acquiring it. Thus, businesses should focus on retention strategies, such as competitive salaries and benefits, career development and learning opportunities, and fostering a positive work environment (Dawson, 2020).

Fostering Collaboration: AI projects necessitate cooperation between diverse teams, including technical and business personnel. A culture that promotes collaboration and cross-functional learning should be encouraged. Regular group meetings, collaborative platforms, and team-building exercises can help nurture this environment.

Leadership: Leaders who understand AI's transformative potential, can strategically guide its implementation, and are capable of navigating AI projects' intricacies are needed. This might involve AI training for existing leaders or recruiting individuals with AI expertise (Bergeret & Malaurent, 2023).

As businesses adapt to rapidly evolving AI technologies, in-house development of AI capabilities allows quick adaptation to changing technologies and market conditions, fostering a culture of continuous learning. Businesses can harness AI's power by investing in talent, creating a conducive work environment, driving innovation, and maintaining a competitive edge.

Change Management

Adopting AI often necessitates considerable adjustments to existing business processes and workflows. Hence, investing in comprehensive change management initiatives is essential to mitigate disruption and facilitate a seamless transition. To successfully implement AI, it is necessary to communicate its benefits to employees, provide training for new processes, and skillfully handle any resistance to change (Antosz, 2021).

Communicating AI Benefits: An essential aspect of change management is transparently communicating the benefits and impacts of AI to all stakeholders, especially employees. It is vital to convey how AI can enhance business performance, improve job efficiency, and create new organizational opportunities. Regular updates, town hall meetings, or dedicated communication channels can be used to keep everyone informed about the AI implementation process and its advantages.

Employee Training and Support: AI may require employees to interact with new technologies and adapt to altered workflows. As a result, businesses need to invest in comprehensive training programs to equip employees with the knowledge and skills required to work effectively with AI systems. This could be done through hands-on workshops and e-learning courses or by hiring external AI experts for specialized training sessions to make things easier. Additionally, we will offer ongoing technical support to help employees overcome challenges while adapting to the new systems (Bergeret & Malaurent, 2023).

Managing Resistance: Change, particularly involving technologies as transformative as AI, can often be met with resistance from employees. Concerns may range from job security to challenges adapting to new technologies. Therefore, companies need to address these concerns proactively. One of the key benefits of AI is its ability to automate repetitive tasks, freeing employees to focus on what they do best: creating business value. If AI implementation leads to changes in job responsibilities, it is essential to provide adequate support, training, and guidance to employees as they adapt to their new roles.

Leadership in Change Management: Leadership plays a crucial role in change management. Leaders need to champion the adoption of AI, set clear expectations, and articulate a clear vision for how AI can drive future growth. They should also listen to employee concerns, show empathy, and respond clearly to their queries (Bergeret & Malaurent, 2023).

In essence, as with any substantial enterprise undertaking, effective change management can significantly enhance the success rate of AI implementations. By ensuring clear communication, providing necessary training, addressing resistance, and showcasing strong leadership, businesses can smoothly integrate AI into their operations, harnessing its full potential to drive business growth (Antosz, 2021).

Governance and Ethics

The introduction of AI projects can precipitate complex ethical dilemmas and governance concerns. These issues encompass algorithmic bias, data privacy, accountability for AI-driven decisions, and more. As a result, organizations need to commit resources to establish comprehensive governance structures and ethical guidelines for AI usage (Cath, 2018).

Setting Up an AI Ethics Committee: Organizations can form an AI ethics committee that brings together members from diverse backgrounds – technology, legal, human resources, and other vital departments – to oversee ethical AI practices. This committee establishes AI usage principles, monitors AI applications for ethical compliance, and makes decisions when moral ambiguity arises. Regular meetings can help to address ongoing concerns and adjust the ethical framework as needed.

Developing a Code of Ethics for AI Use: Businesses should develop a comprehensive code of ethics specifically for AI use. This document would guide the organization's design, development, and deployment of AI applications. It should incorporate fairness, transparency, privacy, and accountability principles. Moreover, this code of ethics should be regularly updated to stay relevant to evolving AI capabilities and societal norms. Compliance and HR training may consider instituting required certifications like those for acceptable use policies.

Implementing AI Monitoring and Auditing Processes: To ensure adherence to established guidelines, companies should implement processes to monitor and audit their AI systems. It is essential to conduct audits to ensure that AI outputs are accurate, there is no algorithmic bias, and data privacy rules are being followed. These audits also reveal any unintended consequences or misuse of AI systems, which can be addressed quickly.

Transparency and Accountability: Organizations must ensure transparency in their AI usage and take responsibility for the decisions made by AI. The ability to elucidate the decision-making process of AI systems, significantly when it impacts individuals who are customers or employees, is essential. Clear lines of responsibility for AI outcomes must also be established to ensure accountability. In addition, companies should provide dedicated channels or reasonable methods to receive anonymous reports of concerns or violations.

As AI's impact on the business environment intensifies, it carries a range of ethical and regulatory considerations. These considerations are of utmost importance for businesses incorporating AI into their operations. Neglecting them can result in severe consequences, including legal penalties, reputational harm, and loss of customer confidence (Cath, 2018).

Regulatory Compliance

Ultimately, businesses must ensure that their AI systems adhere to both existing and forthcoming regulations related to AI and data privacy. Consulting with legal experts and maintaining awareness of regulatory shifts may prove essential (Joshi, 2019).

With AI's pervasiveness growing, regulatory bodies worldwide are working on establishing laws to guide its use. These regulations address AI privacy, transparency, accountability, and bias concerns. Therefore, businesses must stay informed of these regulatory advancements to ensure their AI practices remain compliant. Companies might need to form a dedicated regulatory compliance team or seek legal counsel to ensure responsible AI usage. Compliance is pivotal to avoid penalties and demonstrate commitment to customers, employees, and stakeholders. As AI regulation evolves, businesses must stay current with regulatory changes and ensure compliance with existing and forthcoming laws. Overall, the ethical and regulatory considerations surrounding AI are complex and require careful attention. Businesses must adopt a proactive approach to understanding and addressing these issues as they navigate their AI journey. By doing so, they can responsibly and ethically leverage the potential of AI, providing benefits for all involved stakeholders (Candelon et al., 2021).

By investing in rigorous governance structures and ethical guidelines, businesses can ensure that their AI deployments are effective, efficient but also ethical, and compliant with regulations. AI can fully express its transformative potential by fostering trust amongst stakeholders and protecting the organization's reputation (Candelon et al., 2021).

Partnerships and Collaborations

For many businesses, especially small-to-medium enterprises (SMEs), building a comprehensive in-house AI capability may be neither feasible nor cost-effective. In such scenarios, engaging in strategic partnerships and collaborations with AI technology providers, consulting firms, or academic research institutions can present a more streamlined and efficient approach (Schick, 2023).

Collaboration with AI Technology Providers: Partnerships with established AI technology providers can grant businesses early access to advanced AI tools and platforms. These providers often offer tailored solutions and technical support, simplifying the process of integrating AI into existing business operations (Bergeret & Malaurent, 2023). Furthermore, they can train employees, ensuring the company can utilize these AI tools effectively.

Working with Consulting Firms: AI consulting firms advise businesses to leverage AI to meet their objectives. They can offer expert guidance on selecting, implementing, and managing AI technologies. Additionally, they can assist in navigating the change management process, ensuring a smooth transition as AI is incorporated into business workflows ("AI Partnerships," 2022).

Establishing Ties with Research Institutions: Collaboration with academic or research institutions can offer businesses access to cutting-edge

AI research and innovations. It can also create opportunities for talent acquisition and skill development. This symbiotic relationship can aid institutions by providing real-world applications for their research and businesses by gaining insights into future trends and innovative AI methodologies.

Partnership with Data Providers: AI algorithms need vast data to learn and make accurate predictions. Businesses lacking sufficient internal data can consider partnering with data providers or exchange platforms to gain access to relevant data while respecting privacy regulations.

Investing in AI Startups: Large businesses can strategically invest in AI startups to access innovative AI solutions, while startups can benefit from financial support and business expertise.

In conclusion, strategic partnerships and collaborations can expedite AI implementation and grant businesses access to valuable external resources and expertise. This approach can be particularly beneficial for companies that need more help or expertise to build a comprehensive in-house AI capability, ensuring they can still harness the transformative power of AI (Schick, 2023).

Experimentation and Innovation

As AI evolves, businesses must maintain agility, constantly learning and adapting to new technological advancements. They should devote resources to experimentation and innovation with novel AI technologies and methodologies to stay ahead (Cockburn et al., 2018).

Setting Up an Innovation Lab: Innovation labs provide a dedicated space for experimenting with AI technologies. These labs often act as a sandbox, allowing the exploration of new ideas in a risk-free environment. Labs can be physical to encourage hands-on access, in-person collaboration, or virtually-managed desktops enabling remote access. It can stimulate creativity, leading to the development of novel applications and approaches to business problems. These labs can also be a hub for AI talent, attracting forward-thinking individuals who can drive the company's AI initiatives. Lastly, these labs state a genuine commitment to upskilling the workforce and allowing them to take time out of their regular day-to-day responsibilities.

Participation in Industry Consortia: Joining industry consortia provides a platform for sharing knowledge and insights about AI advancements. It also allows businesses to align their strategies with industry standards and best practices. Participation in these consortia can help companies build networks with AI experts and other firms, fostering collaborative problem-solving and innovation.

Sponsoring Academic Research: Sponsoring academic research in AI can allow businesses to stay abreast of cutting-edge developments. It provides companies access to expertise and innovative thinking while giving academic researchers valuable insights into real-world business challenges. This mutual exchange can often lead to novel AI applications tailored to the sponsoring company's needs.

Investing in Employee Training and Development: Innovation is not merely about technology but also about people. Businesses should invest in ongoing employee training and development programs to foster a culture of innovation. Encouraging employees to gain new skills and knowledge can lead to more innovative applications of AI within the business.

Host Innovation Challenges or Hackathons: Businesses can stimulate innovation by hosting AI-focused hackathons or challenges. These events can bring together employees, partners, customers, and external experts to collaborate on novel AI solutions, fostering an environment of creative thinking and innovation.

By fostering an innovative culture and investing in experimental approaches, businesses can remain at the forefront of AI technology. In this rapidly changing field, an emphasis on continuous learning and experimentation is critical to maintaining a competitive edge and capitalizing on the potential of AI (Cockburn et al., 2018).

Conclusion

As we navigate the complexities of AI in business, it becomes evident that a holistic approach is vital. This chapter has underlined the significance of building robust technical infrastructure, securing, and respecting data, adhering to regulatory compliance, forging strategic partnerships and collaborations, and fostering an environment conducive to experimentation and innovation. It is crucial for businesses not just to deploy AI for its transformative potential but to implement it ethically and responsibly (Bergeret & Malaurent, 2023).

Strategic investments in AI must coincide with building a robust technical framework, comprehensive data privacy and security policies, and rigorous governance structures. Furthermore, an active engagement in strategic collaborations can supplement in-house capabilities and expedite AI implementation, particularly for small-to-medium enterprises. Meanwhile, nurturing an innovative culture within the organization can keep the business at the cutting edge of AI technology, ensuring competitiveness in a rapidly evolving market. Ultimately, the successful and ethical implementation of AI can drive efficiency, profitability, and trust among all stakeholders, unlocking an immense potential for growth and societal impact ("National Institute of Standards and Technology," 2023).

Chapter 7

AI AND CYBERSECURITY: INTERSECTING FRONTIERS

Securing the vast amount of data companies acquire, store, and manage has become a critical issue for businesses, governments, and individuals. Simultaneously, Artificial Intelligence (AI) has seen rapid progress, creating a world that leans heavily on intelligent systems. This presents an opportune time for bad actors to exploit the same capabilities (Raviv, 2021).

AI's role in cybersecurity is twofold. On the one hand, it is a powerful tool for security professionals to detect threats and respond swiftly. On the other hand, it is also a potent tool in the hands of cybercriminals, capable of crafting more sophisticated attacks. This dual nature of AI creates a complex and evolving dynamic.

Cybersecurity in the Age of AI

The increasing reliance on digital platforms and the exponential growth in data have made cybersecurity a strategic priority. Cyber threats have evolved too, becoming more sophisticated and more damaging. The emergence of AI in this environment provides both opportunities and challenges (Anderson, 2023).

AI offers unprecedented capabilities for threat detection, analysis, and response. It can process enormous amounts of data, identify patterns, and make predictions, all at a speed and scale impossible for humans. These capabilities make AI an indispensable tool for modern organizations' security posture. However, as is often the case with powerful technologies, AI's potential can be exploited for malicious purposes (Violino, 2022). With AI, cyber threats can become more intelligent, adaptive, and challenging to detect. This raises the stakes in the cybersecurity game, making it a constant cat-and-mouse chase.

Cybersecurity in the age of AI is a complex equation. It is a balance between leveraging the benefits of AI and managing the risks it brings along. Understanding this dynamic is critical for anyone responsible for data protection and cybersecurity in their organizations.

Role of AI in Enhancing Cybersecurity

AI holds the potential to revolutionize cybersecurity. By applying AI techniques such as machine learning, cybersecurity systems can move beyond traditional rule-based systems. Instead of relying solely on predefined rules to identify threats, AI can learn from the data, identifying patterns and anomalies that might indicate a security breach (Moisset, 2023).

AI can also automate and speed up threat response. AI can promptly implement corrective actions when a threat is detected, minimizing potential damage. Furthermore, AI can predict future threats based on past patterns, making proactive defense possible. Machine learning, a subset of AI, is particularly beneficial in this context. It can process vast volumes of security logs, identify patterns in real time, and evolve its understanding as new data comes in. This capability is instrumental in today's cybersecurity capabilities, where the amount of data to be processed often exceeds human capacity.

However, as powerful as AI is in enhancing cybersecurity, it is important to remember that it is not a silver bullet. It is a tool that can greatly enhance cybersecurity efforts but requires proper implementation, training, and management. A poorly trained AI can produce false positives, miss real threats, or be exploited by clever attackers.

AI-Powered Attacks: The Other Side of the Coin

While AI is a powerful tool for defenders, it is also a potent weapon for attackers. As AI technologies become more accessible, they are increasingly used to conduct more sophisticated, targeted, and impactful cyber-attacks (Moisset, 2023).

One such application of AI in cyber-attacks is in social engineering (Driz, 2023). AI can analyze vast amounts of data from social media and other sources to craft highly personalized and convincing phishing emails. These emails can then be sent to thousands of potential victims, increasing the likelihood of successful attacks.

AI can also automate and scale cyber-attacks. For instance, AI algorithms can conduct repetitive tasks such as password cracking or network scanning much more quickly and effectively than humans. This allows attackers to exploit vulnerabilities more swiftly and evade detection.

Moreover, AI can learn and adapt during an attack. This means it can change its behavior based on the responses of the target system, making the attack more difficult to detect and respond to. This evolving nature of AI-powered attacks makes them particularly challenging for cybersecurity professionals.

Case Studies: AI in Cybersecurity in Action

On the defense side, Darktrace is a cybersecurity company that leverages AI to detect and respond to threats in real-time. Darktrace's AI platform identifies unusual behavior on a company's network that could indicate a cyber threat ("AI Interrupting Cyber-Attacks," 2023). It can even autonomously respond to threats, providing organizations with crucial time to manage the incident. The AI learns from each event, continually updating its understanding of what normal behavior looks like and identifying deviations.

In another instance, Microsoft uses AI to guard its vast cloud infrastructure ("Microsoft AI Defender for Cloud", 2023). Azure Security Center uses machine learning to sift through billions of data points across the network, identifying unusual patterns that could indicate a security breach. It is a feat that would be nearly impossible for human teams to achieve at the scale and speed required.

On the attack side, the cybersecurity team at Check Point Research has unearthed incidents where hackers have co-opted OpenAI's ChatGPT for the creation and refinement of malicious software (Rees, 2023). An alarming case found on a hacking forum involved an individual experimenting with ChatGPT to mimic and refine sophisticated hacking techniques that have been documented in academic research. This highlights the adaptability of cybercriminals and underscores the need for businesses and technology leaders to stay informed and vigilant about evolving threats in the cybersecurity landscape (Rees, 2023).

These cases illustrate the two sides of AI in cybersecurity. They underscore the potential of AI to improve cybersecurity measures and the need for vigilance against AI-powered attacks (Moisset, 2023).

Future Trends for AI and Cybersecurity

As the role of AI in cybersecurity continues to evolve, several trends are worth watching.

First, we can expect AI-powered threat detection and response to become the norm. As cyber threats become more sophisticated, traditional methods will need help to keep up. AI offers the promise of staying one step ahead, continually learning, and adapting to new threats.

Second, the rise of AI in cybersecurity will increase the demand for skilled professionals who can bridge the gap between AI and cybersecurity (Moisset, 2023). Cybersecurity professionals will need to understand AI, and AI professionals will need to understand cybersecurity. This interdisciplinary trend is likely to shape the future job market.

Third, the ethical implications of AI in cybersecurity will come to the forefront. "Issues such as bias in AI algorithms, privacy concerns, and the potential for misuse of AI will require careful consideration ('Impact of AI on Future of Work,' 2023)." These ethical challenges will likely become a key part of the cybersecurity conversation.

However, amid all these trends, one thing is certain. The intersection of AI and cybersecurity will be a key battleground in the ongoing war between cyber attackers and defenders.

Ethical Considerations: AI in Cybersecurity

In the world of AI and cybersecurity, ethics is an increasingly important concern. While AI offers significant advantages in detecting and responding to cyber threats, it also raises ethical questions that businesses need to address (Elkady, 2023). One significant ethical concern is privacy. AI systems often need access to vast amounts of data to function effectively. How can organizations ensure that they respect the privacy of individuals while still leveraging AI for cybersecurity?

Another ethical question relates to the potential misuse of AI. If AI systems fall into the wrong hands, they can be used to perpetrate sophisticated and damaging cyber-attacks. How can organizations ensure that the AI technologies they develop are not used for malicious purposes?

A third ethical concern revolves around the autonomous nature of many AI systems. If an AI system autonomously detects and responds to a cyber threat, who is responsible if something goes wrong? Is it the AI system itself, the people who programmed it, or the organization that deployed it?

Addressing these ethical questions is a complex task. It requires a multidisciplinary approach, involving technical experts, ethicists, legal professionals, and decision-makers. Only through such a collaborative approach can we hope to navigate the ethical demands of AI in cybersecurity.

AI's use in cybersecurity is a rapidly evolving field, and while there are certainly challenges to overcome, the potential benefits of integrating AI into cybersecurity efforts are immense (Driz, 2023). By staying informed and proactively addressing ethical concerns, businesses can leverage AI to enhance their cybersecurity efforts, protecting their valuable digital assets in an increasingly connected world.

Bias in AI Systems and its Impact on Cybersecurity

In AI and cybersecurity, it is important to consider the potential impact of biases in AI systems. AI, including machine learning algorithms, are only as good as the data they are trained on. If this data contains biases, the AI system will likely replicate these biases, leading to potentially skewed or unfair outcomes. In cybersecurity, this could have significant implications.

Consider a scenario where an AI system trained to detect cyber threats is inadvertently biased against certain types of behavior, IP addresses, or geographic locations. This could lead to a disproportionate number of false positives from those areas, draining resources and potentially leading to unjust outcomes. For instance, an innocent activity might be flagged as malicious, causing unnecessary scrutiny or action against a user.

Furthermore, if attackers understand the bias inherent in an AI system, they might exploit it. For example, if an AI system is less likely to flag activity from certain regions, an attacker might route their attack through that region to avoid detection.

Bias in AI systems is a challenging problem to solve, but awareness is the first step. It is crucial for organizations to consider potential biases in the data they use to train their AI systems and implement strategies to mitigate the potential adverse effects.

Cross-Domain AI Applications: Cybersecurity Across Industries

AI in cybersecurity does not confine itself to one industry or sector; it pervades all areas of business and government where data protection is crucial. Its role is particularly pronounced in industries like healthcare, finance, and critical infrastructure, where data sensitivity is high. In healthcare, AI assists in safeguarding patient records, research data, and personal information. It aids in predicting, identifying, and responding to cyber threats, thereby protecting the sanctity of critical health information. Moreover, with a surge in telehealth services and digital health records, the reliance on AI for cybersecurity has escalated.

In the finance sector, another high-stakes arena for cybersecurity, AI has become instrumental. From monitoring transactions to detecting fraudulent patterns, AI provides robust defense mechanisms that ensure financial data integrity. Its ability to swiftly identify anomalies and potential breaches allows for immediate remedial actions, minimizing the risk of financial losses.

Meanwhile, for critical infrastructure, such as energy grids, transportation systems, and other vital services, AI is increasingly adopted to safeguard from cyber-attacks. With smooth operation and public safety at stake, AI enables continuous monitoring and rapid response, ensuring that potential threats are addressed swiftly to prevent service disruptions.

Human-in-the-loop Security

Despite the prowess of AI in managing cybersecurity, the need for human expertise remains indispensable (Moisset, 2023). It is important to understand that AI is a tool, among others, capable of processing vast amounts of data and identifying patterns beyond human capacity. However, nuanced decisions, particularly those with ethical implications or requiring a deep understanding of the context, necessitate human intervention. Humans provide the necessary oversight, discernment, and critical thinking that machines lack.

Moreover, the cybersecurity domain is continually evolving, and threats are becoming more sophisticated. The ability of human professionals to learn, adapt, and innovate is critical in developing strategies that can effectively counter these evolving threats. Hence, an efficient cybersecurity strategy should ideally integrate AI's data-crunching abilities with the discernment and experience of human professionals ("Role of AI in Cybersecurity," 2021).

Furthermore, it is worth noting that the successful deployment of AI in cybersecurity requires human involvement. From data curation and AI training to system monitoring and maintenance, human expertise plays a crucial role at every stage, ensuring that AI performs optimally and safely.

The Role of Open-source Software in AI and Cybersecurity

Open-source software has an important role in AI and cybersecurity. It offers a transparent and community-driven approach that allows researchers and developers to build upon existing frameworks, thereby fostering innovation. The inherent transparency of open-source software can contribute to robust security measures as the code undergoes extensive scrutiny and testing by the community. This broad-based participation can lead to the early identification and remediation of vulnerabilities, making open-source software a viable choice for building secure systems (Elliott, 2022).

However, the openness that makes open-source software so valuable can also pose challenges. The visibility of the source code could potentially aid adversaries in identifying and exploiting vulnerabilities. These adversaries could include not only cybercriminals but also state actors with substantial resources, making the risk even more significant.

Therefore, using open-source software in AI and cybersecurity requires careful handling. It involves balancing the benefits of openness, such as community collaboration and rapid innovation, with potential security risks. Appropriate strategies, like regular software updates and stringent code review processes, can help mitigate these risks while leveraging the potential of open-source software (Korolov, 2022).

AI in Cyber Risk Management and Insurance

With escalating cyber threats, the domain of cyber risk management and insurance has gained prominence. AI, with its capability to predict and identify potential threats, could play an instrumental role in shaping cyber insurance policies. It can help in determining the risk profile of an organization and assessing its vulnerability to different types of cyber threats. This can inform decisions around policy premiums, coverage limits, and other insurance parameters.

In addition to shaping insurance policies, AI can also support proactive risk management. By identifying vulnerabilities and predicting threats, AI can enable organizations to fortify their defenses and preempt cyber-attacks (Moisset, 2023). This proactive approach not only strengthens cybersecurity but can also lead to favorable insurance terms, creating a win-win situation.

Simultaneously, AI can help organizations improve their cybersecurity posture, which can have a positive impact on their insurance standings. It allows organizations to demonstrate their commitment to cybersecurity, potentially leading to reduced premiums and better coverage options. It is an area where AI's predictive capabilities, combined with human-led strategy and decision-making, can bring significant benefits.

Cybersecurity and AI in Post-Quantum Era

The advent of quantum computing promises to bring revolutionary changes in many fields, including cybersecurity. Quantum computers, with their superior processing power, could potentially break cryptographic systems that protect today's internet, posing a significant challenge to AI-driven cybersecurity. However, this does not spell doom for cybersecurity. Instead, it marks the beginning of a new chapter, where old strategies will be revised, and new ones will be devised.

At the same time, the post-quantum era is not all about challenges. The same technology that threatens current cryptographic systems can also lead to new, more robust security measures. Quantum encryption, for instance, promises unbreakable security, offering a glimpse into the potential of quantum technology in enhancing cybersecurity (Driz, 2023).

Therefore, preparing for a post-quantum era involves not just understanding quantum threats but also exploring quantum-based defense mechanisms. It is an emerging frontier where AI could play a crucial role, working in tandem with quantum technology to enhance cybersecurity measures in a rapidly evolving technological environment (Moisset, 2023). As we venture into this new era, it is crucial to stay ahead of the curve, harnessing the potential of AI and quantum technology to secure our digital world.

Conclusion

Artificial Intelligence is undeniably a game-changer in the field of cybersecurity. It has been an ally, bolstering defense capabilities, predicting potential attacks, and rapidly responding to threats. However, it is also a tool that adversaries could wield, becoming a new weapon in their arsenal to launch advanced cyber-attacks.

Navigating this dichotomy is a challenging feat. It demands astute awareness, constant learning, strategic planning, and consistent implementation. Furthermore, the issue of bias in AI systems presents another layer of complexity, necessitating diligent monitoring and evaluation to ensure fairness and efficacy.

However, amidst these challenges, the potential benefits of AI in cybersecurity are compelling. By leveraging AI's capacity for pattern recognition, anomaly detection, and swift response times, we can fortify defenses and create a robust shield against evolving threats.

The rapidly advancing domain of AI and cybersecurity offers immense potential for businesses willing to embrace, adapt, and harness it effectively. It is not just about surviving in the digital age but thriving and leading the way.

Chapter 8

ROLE OF GOVERNMENT IN AI REGULATIONS

AI carries tremendous potential. It promises to make systems more efficient, decisions more informed, and our lives more convenient. However, like all technological innovations, AI has its share of challenges and risks. The breadth and depth of AI's impact necessitate careful oversight to ensure ethical use, safeguard privacy, and maintain security (Aurangzeb, 2023). As the applications and implications of AI continue to grow, it becomes crucial to address the role of government and regulation.

Regulation in this context does not merely refer to the imposition of rules and restrictions. It also involves creating an environment that fosters innovation while protecting societal values. The challenge for governments worldwide lies in striking a balance between these two objectives. It is a tall order, but one that's undeniably essential given AI's transformative potential.

It is worth noting that this is not the first time governments have found themselves at the crossroads of technological innovation and societal welfare. History is replete with examples where government intervention has been instrumental in advancing technological advancements. These past experiences can offer valuable lessons as we navigate the relatively uncharted waters of AI.

Historical Context

The role of government in technological development is not a new concept. Throughout history, government intervention has played an important role in shaping the trajectory of technological advancement. From facilitating early research to funding infrastructure, governments worldwide have fostered innovation.

Take the internet, for example. The genesis of the Internet traces back to the U.S. government's initiative to create a robust, fault-tolerant communication system. The resulting network of networks—ARPANET—was the precursor to the internet we know today. The government's role did not stop at creation; it extended to overseeing the Internet's expansion and managing its transition to commercial use.

Similarly, the role of government in promoting space technology is noteworthy. Public investment and regulation have been vital in propelling advancements in space exploration. The Apollo moon missions, the Mars rovers, and the Hubble Space Telescope—all stand as a testament to what can be achieved when government leads the way.

The emergence of AI as a dominant technology presents a new challenge for governments. Drawing parallels from historical experiences can provide useful insights into how government intervention can facilitate AI's responsible growth.

The Intersection of AI and Public Policy

AI intersects with public policy on multiple fronts. AI technologies, due to their transformative potential, raise questions that are not merely technological but societal as well. From privacy issues to workforce displacement, from the digital divide to the ethics of autonomous decision-making—the range of policy implications is vast and complex.

Privacy is a key concern when it comes to AI. "AI systems often require vast amounts of data to function effectively (Aurangzeb, 2023)." As a result, they can inadvertently infringe on privacy norms. It raises questions about data collection, storage, and usage—areas that public policy needs to address.

Workforce displacement is another policy concern. AI, particularly automation, threatens to displace jobs. While it may also create new jobs, there is no guarantee that those displaced will be equipped to fill these new roles.

The challenge lies in ensuring a just transition and public policy can play a crucial role in managing this change.

AI also risks exacerbating the digital divide. If access to AI technologies remains concentrated among the privileged, it can intensify existing inequities. It is imperative that public policy addresses the issue of equitable access and use of AI.

These are just a few examples of how AI intersects with public policy. As AI continues to evolve, so will its policy implications. It underscores the need for ongoing vigilance and proactive policymaking in this domain.

Case Studies of Government's Role in AI

Governmental interventions have significantly influenced AI development and applications, illustrating the sector's strategic importance. For instance, Singapore launched the national AI Strategy.

In 2019, the city of San Francisco banned the use of facial recognition technology by city departments, citing concerns about civil liberties. It was a groundbreaking decision, underscoring the potential of local governments to shape the scope of AI use within their jurisdictions (Conger et al., 2019).

At the national level, Singapore offers an interesting example. It is one of the first countries to establish a nationwide AI governance framework, called the National AI Strategy, focused on strategic sectors like transport and logistics, indicating the positive role government can play. The model covers key areas such as decision-making, interpretability, and data bias. It aims to foster AI use that is human-centric and ethically aligned. This initiative focused on promoting AI-driven solutions, creating a vibrant AI ecosystem, and developing AI talents and skills.

Another instructive case is the European Union's proposed regulation on AI. It is arguably the most comprehensive regulatory attempt to date, covering a broad spectrum of AI applications. The proposal categorizes AI systems based on their risk level and prescribes corresponding regulatory requirements.

On the contrary, in a more cautionary tale, IBM's Watson for Oncology, which aimed to assist doctors in diagnosing and treating cancer, encountered significant hurdles in countries like South Korea and India. Despite the technology's promise, several reports pointed to a lack of transparency in Watson's decision-making process. The governments and medical associations demanded clarity, revealing how governmental oversight can impact AI's use and reception (Swetlitz & Ross, 2022).

These cases demonstrate that government intervention can take various forms, each with its pros and cons. They underline the need for context-specific strategies that align with local realities and priorities. They also illustrate governments' influence in shaping AI adoption, emphasizing the necessity for proactive and strategic government involvement to ensure AI's beneficial and ethical use.

Need for AI Regulation

Regulation in AI has become crucial for several reasons. AI's capacity to process vast data quantities has raised privacy concerns. For instance, AI applications in healthcare or financial services often involve sensitive personal data. Without stringent regulations, such data could be misused or improperly handled, leading to significant privacy violations (Hon. Donelan, 2023).

Moreover, AI's decision-making process often lacks transparency, which is problematic in sectors like healthcare, criminal justice, or finance, where decisions can have far-reaching consequences. Regulations are necessary to ensure accountability and that AI's decisions are fair and explainable (Hon. Donelan, 2023).

Safety is another critical concern. As AI applications become more intricate and autonomous, the risk of unintended or harmful actions increases. Regulations can help ensure that AI systems are tested rigorously and meet established safety standards (Hon. Donelan, 2023).

Lastly, AI's impact on jobs underscores the need for regulation. While AI can boost productivity, it could also automate many jobs, leading to significant workforce disruption. Policymakers need to mitigate such effects through measures like retraining programs and social safety nets.

Current Regulatory Environment

The current regulatory environment for AI varies significantly across the globe. Several countries have initiated attempts to establish regulatory frameworks, but there is not a globally accepted standard yet.

In the United States, for instance, AI regulation is sector-specific. Different government agencies oversee AI applications within their jurisdictions. The Food and Drug Administration (FDA) regulates AI in healthcare, while the

National Highway Traffic Safety Administration (NHTSA) oversees AI in autonomous vehicles (Felz et al., 2023).

In the European Union (EU), a more holistic approach is pursued. The EU has proposed a comprehensive framework for AI regulation, extending from privacy protections to transparency requirements. The intent is to create a "trustworthy AI" that respects European values and rules (Doyle & Grattirola, 2022).

Asia, meanwhile, presents a mixed picture. Countries like Singapore have introduced comprehensive frameworks for AI governance. China, a leading player in AI, has made significant strides in AI development but faces criticism for its approach to privacy and individual rights (Keane, 2022).

It is evident that AI regulation is a complex, multifaceted challenge that extends beyond national borders. The diversity in current regulatory practices reflects the nascent stage of AI and the distinct socio-cultural and political contexts of different regions.

Perspectives on AI Regulation

The question of how to regulate AI is not just a policy question; it is a societal one. The perspectives on AI regulation are diverse, reflecting a range of concerns, expectations, and visions for the future of AI.

Some argue for a laissez-faire approach, where innovation is not hindered by premature regulation. They posit that regulatory intervention should be minimally invasive, aimed primarily at avoiding harm while encouraging innovation.

Others advocate for a precautionary approach. They argue that the potential risks of AI—such as deep fakes, autonomous weapons, and intrusive surveillance—necessitate strict regulation. This perspective emphasizes protecting societal values and ensuring public trust.

A third viewpoint proposes a collaborative approach, where governments, the private sector, academia, and civil society work together to shape the use and prevalence of AI. This perspective emphasizes dialogue, cooperation, and consensus-building as the means to navigate the complexities of AI regulation.

These differing perspectives are not mutually exclusive. They represent the breadth of considerations that need to be balanced in formulating AI regulation. It is a delicate balance to strike, but an important one in ensuring that AI serves the public good.

Challenges in AI Regulation

Regulating AI presents distinct challenges that require thoughtful responses from policymakers. The primary challenges emerge from the very characteristics that make AI unique and potentially transformative.

AI is inherently complex, making it difficult to understand and oversee. Its decision-making process can be opaque, raising issues of accountability and transparency. This opacity poses a fundamental challenge to regulatory efforts, as regulators must understand AI's functioning to establish effective controls.

AI also evolves rapidly, and the pace of technological advancement can outstrip regulatory processes. Traditional regulatory methods may be ill-equipped to keep up with the speed of AI development. This temporal mismatch poses a significant challenge to ensuring that regulation remains relevant and effective.

Lastly, the impacts of AI are widespread and multifaceted, spanning various sectors and societal domains. This makes it difficult to develop a comprehensive regulatory framework that adequately addresses all potential risks and opportunities.

Despite these challenges, there is a consensus that AI needs regulatory oversight to ensure its responsible use. Regulators will need to be adaptive and forward-thinking to navigate these challenges and develop meaningful, effective regulations.

Future of Government's Role in AI

How the future of the government's role in AI will unfold remain unclear, but several trends can guide our expectations. As AI continues to evolve, venturing into new areas like quantum computing or synthetic biology, governments must adapt regulations to stay relevant and effective. In early May 2023, the Biden administration, for example, held an AI summit with technology leaders and "announced a series of measures to address the challenges of AI, driven by the sudden popularity of tools such as ChatGPT and amid rising concerns about the technology's potential risks for discrimination, misinformation, and privacy (Fung, 2023)."

AI, like many other technologies, does not respect national borders. As AI applications and their impacts become increasingly global, governments need to collaborate more closely. This cooperation could involve harmonizing regulations, coordinating responses to global AI challenges, or jointly developing international AI standards.

Public sentiment is another key influence. As the public becomes more aware of AI and its implications, demands on issues like privacy, transparency, and employment impacts will shape government responses (Fung, 2023). Governments will need to proactively engage with the public, soliciting their input in policymaking processes and addressing their concerns.

Lastly, governments will likely play a more active role in guiding AI development and use. This could involve setting strategic AI priorities, investing in AI research and development, or promoting the ethical use of AI. As the implications of AI become increasingly far-reaching, a more hands-on approach will be necessary to ensure that AI is used for the public good. In the same AI summit, the Biden administration announced plans to introduce policies that shape how federal agencies procure and use AI systems. "The National Science Foundation will also spend $140 million to promote research and development in AI (Fung, 2023)."

As we navigate further into the AI era, the role of governments in shaping the trajectory of AI development will be pivotal. Striking the right balance between enabling innovation and ensuring the ethical, responsible use of AI will be a key task for policymakers worldwide.

AI regulation should be seen not just as a restrictive measure but also as a means to foster trust in AI systems. When people trust AI, they are more likely to use it, unlocking its potential benefits. From a business standpoint, regulatory clarity can provide a stable environment for investment and innovation, fostering the growth of a vibrant AI ecosystem.

Conclusion

In conclusion, governments will need to adopt a balanced and adaptive approach as they tackle AI regulation. A continuous dialogue with a wide range of stakeholders, including technologists, businesses, civil society, and the public, is essential. Although the road to effective AI regulation is complex, it is a necessary journey to ensure a future where AI serves the common good.

Chapter 9

IMPACT OF AI ON THE GLOBAL ECONOMY

Although seemingly a recent player in everyday life, Artificial Intelligence (AI) has been silently revolutionizing our economic systems. Its reach extends beyond mere cost-cutting and efficiency improvements to fundamentally redefining the rules of economic engagement (Bughin et al., 2018).

Simply put, AI is altering how we produce, distribute, and consume goods and services. It accelerates productivity, changes trade dynamics, and even influences fiscal policies. However, its effects are not uniformly distributed. The capabilities and benefits of AI are spreading at varying rates across different countries, leading to potential disparities between developed and developing nations (Furman, 2019).

Economic Impact of AI

Increased productivity is one of the most direct outcomes of AI deployment. For example, companies can leverage AI to automate routine tasks, freeing human resources for more complex and creative endeavors (Bughin et al., 2018). Such a shift can lead to significant productivity gains and economic growth.

AI's influence extends to global trade as well. Traditional trade dynamics are being challenged by AI's transformative potential. For instance, AI-powered supply chain automation is reducing logistical costs and speeding up the delivery process. On another front, the emergence of digital trade, powered by AI tools and platforms, is creating new forms of trade and competition (Furman, 2019).

National Perspective

From a national perspective, AI can boost the nation's economy and drive growth. It can stimulate innovation, productivity, and economic growth by optimizing existing industries and creating new ones. For instance, AI can transform agriculture through precision farming or create entirely new sectors, such as autonomous vehicles and smart homes (Tarswell, 2018).

Moreover, AI can enhance a country's competitiveness on the global stage. AI can help countries gain an edge in the global market by driving innovation and efficiency. This is particularly pertinent for countries seeking to transition from labor-intensive industries to more knowledge-based economies ("AI Policy and National Strategies," 2021).

AI can streamline government services, making them more efficient and accessible. For example, AI-powered chatbots can answer citizens' queries around the clock, reducing the need for manual labor and making public services more citizen-friendly.

However, nations must be cautious about potential pitfalls. There is the risk of widening economic disparities between and within countries. There are also concerns about job displacement, privacy, and security. Thus, while capitalizing on AI's potential, nations must also put in place measures to mitigate these risks.

Regional Perspective

At a regional level, AI's implications can be even more profound. In urban settings, AI can make cities smarter and more livable. From optimizing traffic flow to reducing energy consumption in buildings, AI can make cities more efficient and sustainable (Tarswell, 2018) .

For rural areas, AI can provide a lifeline by improving access to essential services. For instance, AI-powered telemedicine can bring quality healthcare to remote locations, while AI-enhanced e-learning platforms can offer quality education to rural populations ("AI Policy and National Strategies,"2021).

Regionally, AI can also foster economic integration. AI can contribute to regional economic cooperation and development by improving connectivity and facilitating cross-border trade.

However, as with the national perspective, regions must be mindful of the challenges posed by AI. These include socioeconomic disparities, job displacement, and a host of ethical and legal issues. It is crucial to balance harnessing AI's benefits and managing its risks.

AI Divide: Developed vs. Developing Countries

AI holds considerable promise, but the benefits it offers are unevenly distributed globally. Developed nations, equipped with superior digital infrastructure and ample resources, are surging ahead in the AI race (Hupfer, 2019). These countries are effectively utilizing AI technologies to bolster their economic prowess, a development that's accentuating the "AI divide".

Developing countries, on the other hand, face considerable challenges in harnessing AI's full potential (Bughin et al., 2018). These range from lacking resources and adequate digital infrastructure to a dearth of skills and policy frameworks needed to navigate the AI landscape. This disparity in AI capabilities between developed and developing countries could result in the latter falling further behind in global economic competition.

The implications of this divide extend beyond economic parameters. As AI becomes increasingly integral to societal functions, those nations lagging in AI adoption may witness a widening digital gap among their population, exacerbating social inequality. The path to bridging this divide is complex, but the urgency to do so is palpable.

In addressing this issue, international cooperation will be key. Wealthier nations and global institutions must play a pivotal role in enabling developing countries to tap into AI's potential. This could involve facilitating technology transfer, fostering skill development, and providing policy guidance (Hupfer, 2019).

Role of AI in Shaping Monetary and Fiscal Policies

AI, with its ability to analyze and interpret vast quantities of data, is becoming increasingly influential in shaping economic policies. In monetary policy, AI could enable central banks to forecast inflation trends more accurately, informing policy decisions related to interest rates (Patel et al., 2022).

AI could also provide valuable insights for fiscal policy-making. For instance, it could assist in managing public debt by enabling better risk assessment and forecasting. AI could also help governments optimize their revenue collection strategies by improving tax fraud detection.

Moreover, AI's ability to model complex economic scenarios could guide governments in designing stimulus packages and social welfare programs. With AI, policymakers could assess the likely impact of such initiatives more accurately, leading to more effective policy interventions. However, there is always a risk of over-reliance on these models, especially with the inherent risks of data quality and bias (Patel et al., 2022).

It is important, however, to acknowledge the limitations and risks associated with AI's role in economic policy-making. Ensuring the quality of data, maintaining transparency in AI models, and addressing the risk of automation bias in decision-making are among the challenges that need to be addressed as AI becomes a critical tool in economic governance (Patel et al., 2022).

AI and the Transformation of Economic Sectors

AI is not just reshaping traditional economic sectors; it is also giving birth to entirely new ones. In manufacturing, AI-driven automation is heralding a new era of precision and efficiency. In agriculture, AI tools are improving crop yields and reducing waste, promoting a more sustainable form of farming (Qureshi, 2022).

In the service sector, AI is transforming customer experiences by enabling personalized interactions. From AI-powered chatbots addressing customer queries to recommendation algorithms offering tailored product suggestions, AI is making customer service more efficient and enjoyable (Qureshi, 2022).

Beyond transforming existing sectors, AI is also spurring the growth of new industries. The autonomous vehicles sector, for example, is flourishing thanks to AI technologies. Similarly, the data analytics industry is witnessing explosive growth as businesses strive to make sense of the voluminous data they collect.

However, these transformations also bring with them certain challenges. For instance, the rise of AI may lead to job displacement in certain sectors. Similarly, the rapid pace of change can cause regulatory and ethical dilemmas. As we navigate this transformation, we must strive to balance the benefits of AI with the potential risks and disruptions.

Case Studies

The sweeping influence of AI on the global economy is undeniable. One prime example is AI's potential to redefine productivity and Gross Domestic Product (GDP) growth. According to a study by PwC, AI could contribute a staggering $15.7 trillion to the global economy by 2030 (Bughin et al., 2018). This monumental figure comprises an anticipated $6.6 trillion boost from increased productivity and an additional $9.1 trillion from consumption-side effects (Bughin et al., 2018).

These gains in productivity stem from the capacity of AI to augment and automate tasks and roles, thereby enhancing labor efficiency1. Furthermore, product enhancements, driven by AI's ability to deliver greater product variety, personalization, attractiveness, and affordability, are projected to contribute 45% of total economic gains by 2030 (Bughin et al., 2018). This transformative potential of AI is not exclusive to developed economies. From a macroeconomic perspective, there are opportunities for emerging markets to leapfrog their more developed counterparts, indicating that the effects of AI are far-reaching and inclusive (Bughin et al., 2018).

AI's role in augmenting productivity and fostering trade opportunities presents another compelling example. By 2030, AI could bolster global output by around 16% or $13 trillion, effectively doubling annual economic growth rates in developed economies (Meltzer, 2022). As AI optimizes productivity, firm competitiveness escalates, creating new prospects for international trade and expanding the share of services in this domain. However, it is crucial to note that the diffusion of AI can lead to regulatory heterogeneity that could impede trade in AI products, underscoring the importance of international coordination in AI regulation (Meltzer, 2022).

Nonetheless, the evolution of AI is likely to present economic and social transition costs, such as rising income inequality and job losses. To mitigate these potential downsides, governments worldwide are implementing frameworks and policies to build domestic AI capabilities, ensuring the transition into an AI-driven economy is well-managed and inclusive.

Innovation Ecosystem

AI's surge has transformed global innovation ecosystems by revolutionizing research and development activities. It is speeding up discovery cycles, enabling the rapid prototyping of ideas, and fostering a culture of fast-fail experimentation. Machine learning algorithms, for instance, are increasingly used to analyze patterns in complex data sets, offering novel insights that drive groundbreaking research across industries. As such, the frontiers of human knowledge are expanding at an unprecedented rate (West & Allen, 2023).

In the startup domain, AI-enabled ventures are coming to the fore, receiving substantial interest from industry veterans and investors. These innovative startups, with their AI-powered solutions, are transforming traditional business models, impacting sectors as diverse as healthcare, finance, retail, and more. AI is no longer just an accessory technology but has become the core value proposition of many emerging enterprises.

Venture capital investments are leaning heavily towards AI-driven businesses, reflecting the high-potential return on investment these technologies promise. This financial push is driving AI innovation to new heights, enabling the development of increasingly sophisticated applications. Meanwhile, increasing AI-related patents underscores the mounting business interest in AI and the desire to secure market advantages through unique AI innovations (Wiggers, 2023).

Economic Theories and AI

AI's emergence challenges existing economic theories, particularly those concerning labor market dynamics. Classical economic theories often hinge on the assumption that labor is a fungible and uniform commodity. However, the advent of AI has begun to disrupt this notion, forcing economists to reconsider fundamental assumptions (Engler, 2022).

In an AI-driven economy, the shift from manual and routine jobs to roles requiring complex cognition and emotional intelligence is evident. This shift does not align with the classic labor market theory that argues for a largely stable and predictable job market. The rise of AI demands an updated economic framework, acknowledging the unique labor dynamics that AI and automation bring to the fore.

Moreover, AI influences supply and demand mechanics, another core component of classical economics. On the supply side, AI technologies, acting as non-human agents, produce goods and services at unimaginable scales. On the demand side, AI's predictive capabilities enable more precise forecasting and targeting, altering consumption patterns. These changes challenge traditional supply and demand theories, requiring new economic models for the AI age (Engler, 2022).

AI, Sustainability, and the Green Economy

The intersection of AI and sustainability is a burgeoning field of interest. AI has the potential to contribute significantly to the green economy, both by improving efficiency and by promoting more sustainable practices (Gast, 2022).

In waste management, AI can streamline recycling processes and help identify new uses for waste materials, contributing to a circular economy. Similarly, AI can improve crop yield predictions in agriculture, leading to more efficient use of resources and less waste.

AI's ability to analyze and interpret vast datasets can also help us better understand environmental trends and predict future changes. This can inform policy-making and help us design more effective responses to environmental challenges (Gast, 2022).

Nevertheless, it is imperative to be mindful of the potential downsides. The proliferation of AI technologies also means a surge in energy consumption, posing challenges for sustainability. As we harness AI's potential for the green economy, we must strive to mitigate these negative impacts.

Trade Relations

AI substantially impacts global trade, ushering in changes that reshape relationships between nations. AI technologies offer capabilities that can streamline supply chain processes, enhance production efficiency, and transform service delivery. These advancements affect international trade dynamics, as countries that successfully harness AI can potentially gain significant competitive advantages (Meltzer, 2022).

AI can also exacerbate existing trade tensions. As countries vie for AI dominance, the prospect of "data nationalism" – where nations enact policies to protect and control their data – becomes a genuine concern. This poses challenges for international trade norms, as data is a crucial input for AI. The risk of a fragmented global data regime, punctuated by protectionist data policies, could disrupt global trade relations and necessitate new agreements or revisions to existing ones.

Lastly, AI's role in automating production can alter countries' comparative advantages. Traditional labor-abundant countries might find their advantages diminished as AI and automation decrease the labor intensity of production processes. This shift can disrupt existing trade patterns and force a reconsideration of international trade strategies (Goldfarb & Trefler, 2018).

Socio-Economic Challenges and Risks

AI's rise is not without challenges and risks. There is a growing concern about how AI could exacerbate wealth and income inequality. As AI continues to drive productivity gains, ensuring that these gains are equitably distributed is crucial. A scenario where a disproportionate share of the benefits flows to the wealthier segments could lead to increased socioeconomic disparities (Marr, 2023).

Moreover, the fear of job displacement is looming large. While AI is expected to create new jobs, it might also render many existing jobs obsolete. This could lead to widespread job insecurity and social unrest.

Another challenge lies in managing the societal transition necessitated by AI's rise. As AI continues to permeate various aspects of our lives, it is imperative to prepare our societies for this change. This includes rethinking our education systems to equip future generations with the skills needed to thrive in an AI-driven world.

Finally, while the AI race is often framed as a competition, it is crucial to foster a collaborative approach. As nations, businesses, and individuals, we all have a stake in the future shaped by AI. Ensuring that this future is inclusive, equitable, and sustainable should be a shared goal.

AI's Impact on Jobs and Skillsets

Integrating AI into business processes raises concerns about job displacement and the changing demands for skills. While AI has the potential to automate specific tasks, history has shown that technological advancements typically lead to the creation of new jobs and opportunities, albeit with a shift in required skill sets (Stahl, 2022).

To prepare for the future of work, organizations must invest in reskilling and upskilling their workforce. This involves training and development programs that enable employees to adapt to the evolving work environment and acquire new skills. By investing in their employee's professional growth, organizations can ensure that they have a skilled workforce that is ready to embrace the opportunities presented by AI (Stahl, 2022).

Collaboration between humans and AI is also crucial in the future of work. Rather than viewing AI as a threat to jobs, organizations should promote collaboration between human employees and AI systems. AI can augment human capabilities, enhance productivity, and assist in decision-making. By fostering a collaborative environment where humans and AI work together synergistically, organizations can maximize the potential of both.

Creating a culture of continuous learning is essential for preparing the workforce for the future of work. Employees must be adaptable and open to learning new skills in a rapidly changing technological world. Organizations should create an environment encouraging ongoing learning and development, providing resources and opportunities for employees to stay relevant and continuously upskill.

By embracing these strategies, organizations can navigate the evolving state of work and leverage the power of AI to drive growth and innovation. Rather than fearing job displacement, they can position themselves to harness AI's opportunities and create a future of work characterized by collaboration, continuous learning, and human-AI synergy.

Ensuring Fair Compensation in the Age of AI

As AI continues to integrate into the workplace, ensuring fair compensation for employees is crucial. This includes addressing the issue of "shadow work," where AI systems are involved in tasks, but the contributions of human workers are overlooked or undervalued.

To ensure fair compensation, businesses must recognize and value the contributions of employees engaged in shadow work. It involves explicitly incorporating such tasks into job descriptions and performance evaluations, ensuring they are accounted for when determining compensation. By acknowledging the human effort involved in training AI systems, correcting errors, or making decisions based on AI outputs, organizations can ensure that employees receive appropriate recognition and compensation for their work.

In addition, businesses can consider implementing profit-sharing or ownership models for AI technologies. This means that employees who contribute to the development or operation of AI systems share in the financial returns generated by these technologies. By linking compensation to the success and impact of AI systems, organizations incentivize employees to actively engage with AI and align their efforts with the business's overall success.

At its core, fair compensation in the age of AI involves recognizing and valuing the human labor that enables and complements AI. It requires organizations to proactively address the issue of shadow work, ensuring that employees are kept from being overshadowed or replaced by AI systems. By approaching this issue ethically and transparently, businesses can attract and retain top talent, promote employee satisfaction, and create a sustainable and equitable work environment in an AI-driven world.

Ensuring fair compensation in the age of AI requires recognizing the contributions of employees engaged in shadow work, incorporating their efforts into compensation decisions, and considering profit-sharing or ownership models for AI technologies. By valuing the human labor behind AI and implementing fair compensation practices, organizations can foster a supportive and inclusive work environment, driving sustainable success in the era of AI.

Conclusion

AI is undeniably poised to reshape the global economy. The potential for transformative change is immense, but so are the challenges and risks. As we increasingly integrate AI into economic systems, it is vital to ensure that this is done in a manner that promotes economic growth, social equity, sustainability, and collective well-being.

Chapter 10

WORKFORCE TRANSFORMATION IN THE AI ERA

Navigating Professional Shifts and Advancements Through AI Innovations

The rapid evolution of artificial intelligence (AI) presents businesses with both opportunities and challenges. This chapter covers the implications of AI for the workforce and skillsets, examines how AI can enhance productivity and efficiency, explores the need for accountability in AI-aided workplaces, and discusses the complexities of fair compensation in the era of AI. By understanding these facets of AI integration, business and technology leaders can better navigate the shifts and advancements AI brings to the workplace (Gandzeychuk, 2023).

Enhancing Productivity and Efficiency with AI: Empowering the Workforce

AI has the potential to enhance productivity and efficiency in the workplace significantly. Rather than solely focusing on task automation, AI can augment human capabilities, enabling workers to perform their functions more effectively (Basu, 2023). By leveraging AI technologies, businesses can empower their workforce and achieve improved business outcomes.

One way AI enhances productivity is through task automation. AI-powered solutions can automate repetitive and mundane tasks, freeing up valuable time for employees to focus on more complex, creative, and strategic work. For example, AI-powered chatbots can handle routine customer queries, allowing human employees to concentrate on providing personalized and high-value customer interactions (Aidude, 2023). Similarly, AI algorithms can automate data entry and processing tasks, enabling employees to devote time to data analysis and decision-making (Aidude, 2023).

AI also plays a crucial role in decision-making processes. By analyzing vast amounts of data, AI algorithms can identify patterns, trends, and insights that may be difficult for humans to detect. This data-driven decision support enables workers to make informed decisions more quickly and accurately. Whether in financial forecasting, supply chain optimization, or customer segmentation, AI-powered analytics can provide valuable insights contributing to better decision-making.

Moreover, AI can improve communication and collaboration within teams. AI-powered tools can streamline and enhance various aspects of teamwork, such as managing and scheduling meetings, facilitating real-time language translation, and even identifying potential misunderstandings or conflicts within a team. By leveraging AI in these areas, employees can communicate more efficiently, collaborate seamlessly, and overcome language barriers, improving productivity and teamwork.

When implemented effectively, AI can be a powerful tool for enhancing worker productivity and efficiency. Rather than replacing humans, AI augments human abilities and frees time for more valuable tasks ("Future of AI," 2023). By automating repetitive tasks, supporting decision-making processes, and facilitating communication and collaboration, AI empowers the workforce to focus on tasks that require creativity, critical thinking, and problem-solving skills (Gandzeichuk, 2023). This leads to a more engaged and effective workforce, ultimately driving business success.

Maintaining Accountability and Authenticity in AI-Aided Workplaces

As AI tools become more sophisticated and accessible, it is essential to establish guidelines to ensure that the work claimed by employees is indeed their own and not delegated to AI tools without employers' knowledge. Maintaining authenticity and accountability in AI-aided workplaces goes beyond academic plagiarism; it extends to professional responsibility and integrity.

One of the critical measures to maintain authenticity is to establish transparent use of AI tools. Employers can set policies requiring employees to disclose when and how they use AI. This approach ensures that the use of AI tools is transparent and that any work completed with the aid of AI is appropriately attributed. By fostering transparency, organizations can uphold the integrity of work produced in AI-aided environments.

Implementing an AI use policy is another crucial step. This policy can outline the acceptable use of AI in the workplace, addressing issues such as when and how AI can be used, how its use should be documented, and how the benefits of AI should be shared. By establishing clear guidelines, organizations can ensure that AI is used responsibly and ethically, preventing potential misuse or misrepresentation of AI-aided work (Basu, 2023).

Education and training play a vital role in promoting the ethical use of AI. Employers need to provide comprehensive training programs that bring awareness to the ethical implications of AI use, including educating employees about issues such as data privacy, bias in algorithms, and the importance of honesty and integrity when utilizing AI tools. By promoting ethical awareness, organizations can cultivate a sense of responsibility and accountability among employees, ensuring that AI is used consistently with professional standards (Basu, 2023).

Regular auditing mechanisms are essential to maintaining accountability in AI-aided workplaces. Employers can periodically review the AI tools employees use and the work produced using these tools. Audits help ensure compliance with policies and guidelines, allowing organizations to identify potential issues or discrepancies. Through regular audits, organizations can reinforce the importance of maintaining accountability and authenticity in using AI.

In conclusion, the responsible use of AI in the workplace requires explicit guidelines, ongoing education, and monitoring systems. Employers can foster a culture of transparency and accountability by establishing transparent use policies, providing education, and training on ethical AI use, and implementing auditing mechanisms. This ensures that AI is used to enhance human work rather than to replace or misrepresent it, upholding integrity and maintaining trust in AI-aided workplaces (Basu, 2023).

Changing Nature of Roles and Professions

As AI permeates the workplace, the dynamics of traditional roles are transforming, and entirely new positions are emerging. This shift requires a fresh look at career paths, skills, and job descriptions.

Even traditional roles, which may seem distant from technological influence, are undergoing changes. For instance, accountants find that AI can handle repetitive tasks such as data entry or basic calculations. This advancement allows them to focus more on strategic advising and decision-making roles. Similarly, healthcare professionals are experiencing AI's growing involvement in diagnostics, which enhances their expertise and allows for more personalized patient care.

Emerging roles, such as AI ethicists who ensure AI systems uphold the ethical values of their organizations and society, are becoming more prevalent. Demand for data scientists and AI trainers is increasing as AI models' development, implementation, and fine-tuning become more integral in businesses. These roles, nonexistent a few decades ago, exemplify the transformative effect of AI on the job market.

A shift towards "prompt engineering" is noticeable even within AI development. With the rise of large language models, the ability to construct effective prompts that guide the model's output is becoming an increasingly valued skill. This ability extends beyond mere technical capability, requiring understanding context, crafting relevant examples, and defining clear tasks for the model.

The evolving nature of professions due to AI necessitates adaptability and proactiveness from businesses and individuals. Lifelong learning and skills enhancement are becoming essential in this rapidly changing environment (Amir, 2023). At the same time, it is crucial for educational institutions and organizations to provide resources and training opportunities. This approach will align the workforce's capabilities with the demands of an AI-integrated world. Embracing these changes ensures a smooth transition into a world of possibilities and opportunities powered by AI.

Gig Workers

Rise and Challenges of the Gig Economy

Over the last decade, the gig economy has grown exponentially, transforming traditional employment models. Technology advancements, the desire for flexible work, and economic shifts have fueled this growth. Workers are attracted by the freedom to choose when and where they work and the ability to diversify their income sources. For businesses, the gig economy provides access to a vast, flexible workforce, reducing fixed labor costs and allowing them to scale up or down as needed (Amir, 2023).

However, the gig economy also presents its own set of challenges. Many gig workers face job insecurity, volatile income, and a lack of benefits and protections typically associated with traditional employment (Amir, 2023).

AI's Role and Impact on the Gig Economy

AI has become a powerful tool in the gig economy, underpinning many digital platforms connecting gig workers with opportunities. Uber, Lyft, and Upwork, among others, use AI to match workers with jobs, predict demand, and set dynamic pricing.

AI is also used to manage and support gig workers. For instance, chatbots handle queries, and AI systems track work, collect payments, and even predict workers' behavior. However, this has raised concerns about surveillance and control. Critics argue that these platforms exert significant control over workers, blurring the line between independent contractors and employees.

The impact of AI on job opportunities in the gig economy is a complex issue ("The Future of AI," 2023). On the one hand, automation threatens certain types of gig work; on the other, it may create new opportunities for gig work, particularly in areas like AI training and verification.

Empowering Gig Workers through AI

Despite these challenges, AI has the potential to empower gig workers significantly. AI tools and platforms can help gig workers find jobs, negotiate better rates, develop their skills, manage work, manage financial planning, and potentially complete their tasks faster.

For instance, AI can level the playing field by giving gig workers access to data and insights usually reserved for large businesses. By analyzing market demand, competition, and pricing, AI tools can help gig workers make informed decisions about where to work, when to work, and what to charge.

AI-powered learning platforms can provide personalized, on-demand skill development, helping gig workers stay competitive in the changing marketplace.

Future of the Gig Economy and AI

Several trends will shape the gig economy's future and AI's role. One such trend is the personalization of work. Just as AI can personalize consumer experiences, it can also create personalized work experiences, matching gig workers with opportunities that fit their skills, interests, values, and life circumstances. For example, "Uber recently filed a patent application for a system that would leverage AI and machine learning to match potential riders and drivers based on predictions of customers' tendencies. The system forecasts customers' potential ride requests by feeding "context data" contained in Uber user profiles, such as usage history or location data, into a learning model that uses AI to find patterns in users' habits... The model then determines whether the user should be considered for "pre-request matching" by which a driver can relocate to an optimal location to reduce wait times in the event the user requests a ride (Revell, 2023)."

Data privacy and ethics in AI will become increasingly important. As gig platforms collect vast amounts of data about workers, there are growing concerns about how this data is used and protected. Ethical AI practices will be critical, and regulation may play a role in ensuring these practices (Basu, 2023).

These trends present opportunities and challenges for gig workers, businesses, and society (Amir, 2023). While AI has the potential to enhance the gig economy significantly, careful thought and action will be needed to ensure that it is used in a way that benefits all stakeholders.

Researchers

AI is transforming the field of research across disciplines. It is not just about automating repetitive tasks but rather about augmenting human capabilities and accelerating the research process. As large volumes of data become increasingly central to research in medicine, physics, social sciences, and more, AI can help researchers make sense of this information. Machine learning algorithms can comb through large datasets to identify patterns and correlations, revealing insights that would take human researchers far longer to discover.

For instance, in biomedical research, AI is used to understand genetic data and predict how different genes interact, leading to significant advances in understanding diseases and developing new treatments. Similarly, in social sciences, AI algorithms can process and analyze large amounts of text data, such as social media posts or news articles, to study societal trends and human behavior.

Furthermore, AI tools can automate the literature review process, a time-consuming task for researchers. By rapidly scanning and synthesizing thousands of articles, AI can provide researchers with relevant information and insights, allowing them to stay up to date with the latest developments in their field.

However, while AI brings significant advantages to research, it also introduces new challenges ("The Future of AI," 2023). Aside from generative AI capabilities, using AI requires data science and programming skills, something not all researchers possess. There is a risk that AI could widen the gap between researchers with these skills and those who do not. Moreover, using AI raises questions about the validity and reproducibility of research findings. If AI algorithms are used to identify patterns in data, it is important that these algorithms are transparent and that other researchers can replicate their results.

Another challenge is the veracity and reliability of the content generated. In an ongoing development at the time of the final touches of this book, "a man named Roberto Mata sued the airline Avianca, claiming he was injured when a metal serving cart struck his knee during a flight to Kennedy International Airport in New York. When Avianca asked a Manhattan federal judge to dismiss the case, Mr. Mata's lawyers objected, submitting a 10-page brief that cited more than half a dozen relevant court decisions. [However,] no one could find the decisions, or the quotations cited and summarized in the brief because ChatGPT had invented everything. The lawyer who created the brief admitted in an affidavit that he had used the ChatGPT to do his legal research," "a source that has revealed itself to be unreliable. Judge Castel said in an order that he had been presented with "an unprecedented circumstance (Weiser, 2023)," a legal submission replete with "bogus judicial decisions, with bogus quotes and bogus internal citations (Weiser, 2023)." The lawyer is currently facing potential legal sanctions.

Despite these challenges, it is clear that AI can significantly enhance researchers' capabilities, speeding up the research process and enabling discoveries. The key will be to ensure that researchers are equipped with the skills and knowledge to make the most of these tools and that appropriate guidelines are in place to ensure the responsible use of AI in research.

Editors

AI is changing the game for editors as well. Machine learning algorithms can now perform tasks like spell-checking, grammar-checking, and even checking for writing style and tone. Some AI tools can suggest rewrites for clumsy sentences and check for clarity issues or biased language. This allows editors to focus more on the content and structure of the text rather than getting bogged down in the mechanics of language. The author, for instance, used Grammarly, an online editing tool with AI capabilities, for the editing of the book, saving significant time and cost.

AI tools can also help with fact-checking, a critical aspect of an editor's job. By quickly cross-referencing information across multiple sources, AI can flag potential inaccuracies for human review. This does not replace the need for human judgment but can make the fact-checking process more efficient.

However, as with researchers, the advent of AI tools in editing poses challenges. The automated editing tools could be more foolproof and sometimes miss language nuances that a human editor would catch. Dependence on AI tools could lead to a decrease in human editing skills. Furthermore, data privacy issues arise as these tools often require access to edited text.

Nonetheless, AI has immense potential to assist editors in their work, helping them to produce higher-quality, more accurate content. As these tools evolve and improve, they will likely become an increasingly important part of an editor's toolkit.

Graphic Artists

AI is also making waves in the creative fields, including graphic design. Machine learning algorithms can now suggest design elements based on trends and past successful designs. AI tools can automate tasks like resizing images for different platforms, saving much time for graphic artists. AI can even create designs from scratch based on specific user-provided parameters.

However, perhaps the most significant impact of AI on graphic design is its potential to democratize design. AI-powered design tools can make it easier for non-designers to create professional-looking designs, leveling the playing field and opening up the world of graphic design to more people. The author, for example, used Midjourney, through a number of prompts to refine concepts, to generate the cover for the book.

However, this also raises concerns for professional graphic artists. If anyone can create a decent design using AI tools, will the demand for professional graphic designers decrease? Moreover, while AI can create designs based on parameters and past trends, it lacks the human touch, the ability to understand and convey emotion, and to think outside the box.

Despite these concerns, AI has the potential to be a powerful tool for graphic artists, enabling them to work more efficiently and effectively and to create more impactful designs. By combining the capabilities of AI with human creativity and emotion, graphic artists can push the boundaries of what is possible in design.

Programmers and Software Engineers

Result of AI on Productivity

Their impact on productivity strongly underscores the value proposition of AI tools in software development. This claim is supported by a rigorous study involving GitHub Copilot, an AI pair programmer. In this controlled experiment, software developers were assigned to implement an HTTP server in JavaScript as quickly as possible. The group that leveraged the capabilities of GitHub Copilot outperformed the control group, completing the task 55.8% faster. This significant productivity gain demonstrates the potential of AI tools to supercharge the efficiency of software development processes (Gandzeichuk, 2023). Another interesting aspect of this study was that the AI tools proved particularly beneficial for developers who are either less experienced, older, or devote many hours per day to programming. These findings underscore the democratizing potential of AI tools in software development, making it easier for a broader range of individuals to transition into software development careers.

Beyond academic research, real-world testimonials of developers further affirm the productivity benefits of AI tools. For example, Jonathan Burket, a senior engineering manager at Duolingo Inc., reported a 25% improvement in his efficiency attributed to using GitHub Copilot. This highlights how AI tools can bring tangible benefits to everyday programming tasks, making developers more effective and efficient.

AI's Transformation of Software Engineering

AI is creating a paradigm shift in coding and development activities. The advent of AI-powered tools, including ChatGPT developed by OpenAI and GitHub Copilot, is a testament to this rapid transformation. These tools have introduced new dimensions to programming, enhancing productivity and reshaping traditional software engineering processes. The AI-driven transformation is not only streamlining tasks but also driving a shift in the roles and tasks of software developers.

AI in the Software Development Process

In the current software development process, AI is not just an auxiliary tool; it can become an integral part of various stages:

Requirements Gathering and Testing: The precision and detail necessary in this phase can be significantly enhanced with AI. For instance, OpenAI Codex, when used in conjunction with Selenium, can assist a business analyst and a QA engineer in defining all necessary user stories for particular use cases (Gandzeichuk, 2023). Additionally, it can generate auto-tests covering all possible test cases, ensuring thorough and accurate software testing.

UI/UX Design: The design phase of software development, which is traditionally a highly human-centric task, can also benefit from AI. With the advent of multimodal AI capabilities, such as those exhibited by ChatGPT-4, design specialists can build user interfaces and create customer journeys more effectively (Gandzeichuk, 2023). This not only enhances the quality of design but also speeds up the design process.

Coding: AI has a significant role in the coding phase of software development. Tools like Bing AI can generate code snippets based on given prompts, aiding in rapid prototyping and iterating on different ideas (Gandzeichuk, 2023). This facilitates faster development cycles and innovative solutions. With AI taking up substantial parts of the coding process, the role of senior engineers will likely shift towards verifying and polishing the code generated by AI tools (Gandzeichuk, 2023). This allows for better utilization of their expertise and experience.

Unit Testing: AI tools can perfectly fit this task since tests are typically automated (Gandzeichuk, 2023). For example, CodeWhisperer has demonstrated excellent results in automating unit tests, ensuring the code behaves as expected under various conditions.

Future of the Software Engineer

Contrary to the idea that AI might replace software engineers, "many experts, including Alan Fern, a professor of computer science and executive director of AI research at Oregon State University's College of Engineering, believe that automation tools improve the efficiency of highly skilled developers (Gandzeichuk, 2023)." They help with repetitive tasks, allowing developers to focus on more complex problems. Tools like ChatGPT, Bing AI, Copilot, Tabnine, and Amazon CodeWhisperer are seen as the future of software development rather than as replacements for developers. Instead, they are expected to expedite the pace of modern software development, promote experimentation, and transform the current software engineering funnel (Gandzeichuk, 2023).

AI-powered tools can enhance various stages of the software development process. For instance, they can assist in defining user stories and generating auto-tests, building user interfaces, suggesting relevant services from public cloud providers, writing code, automating unit tests, developing API integrations, and verifying deployments. AI can also assist during the post-deployment phase, flagging errors and uncovering abnormalities by analyzing system logs (Gandzeichuk, 2023). Thus, they help improve the efficiency of the development process, potentially leading to higher-quality software products delivered in less time (Gandzeichuk, 2023).

The software engineering process might transform into two distinct stages in the future — the creative stage and the delivery stage. The creative stage would involve more human collaboration with AI, while the delivery stage would rely more on AI (Gandzeichuk, 2023). The result of this transformation could be more efficient and accurate software development processes, with engineers being able to focus on more complex and creative problem-solving (Gandzeichuk, 2023). Utilizing AI-powered tools can significantly improve the efficiency of software development processes, as observed by Jonathan Burket, a senior engineering manager at Duolingo Inc., who reported that Copilot makes him 25% more efficient (Gandzeichuk, 2023).

"Organizations investing in creating custom software may find automating repetitive tasks through AI technology a potential growth point. This could lead to better quality end-products and quicker turnaround times, making it a promising venture to explore (Gandzeichuk, 2023)."

Rise of Prompt Engineers

Prompt engineering has recently emerged as a valuable skill in AI. In the context of generative AI, which are systems that generate new content, the ability to craft effective prompts is critical. Prompts are the instructions or queries given to an AI system, and they play a crucial role in determining the kind of responses or results the AI will produce.

Prompt engineering uses context, examples, and tasks to guide the AI to provide the desired outcomes. It is about finding the right way to ask the AI system to do something. For instance, how do you ask an AI to write a poem, summarize a complex report, or generate a list of ideas? The way you frame your prompt will significantly impact the AI's output.

The importance of this skill is reflected in the rise of communities on platforms like Twitter, where individuals share tips and guides on how to write effective prompts for AI. These communities are proliferating, signaling the value and interest in prompt engineering.

Just as data analysis or coding skills have become prerequisites for operating in a digitally-driven environment, prompt engineering is steadily establishing itself as a crucial skill for interacting effectively with generative AI systems to achieve results that are more precise and beneficial. As AI grows more sophisticated and is tasked with increasingly complex functions, the significance of prompt engineering is only likely to increase. Hence, awareness and skill development in prompt engineering should be considered a strategic move for any organization or individual in the technology space looking to stay at the forefront of AI advancements.

Conclusion

In each of these professions, AI presents both opportunities and challenges. The key will be to harness the advantages of AI while also addressing the challenges, ensuring that these tools are used to enhance human capabilities rather than replace them. This will require ongoing dialogue and collaboration between AI developers, professionals, and policymakers.

Chapter 11

AI IN HEALTHCARE: CURRENT APPLICATIONS AND FUTURE POTENTIAL

AI applications in healthcare are as diverse as they are significant. Machine learning is helping in diagnostics, where AI algorithms are increasingly assisting doctors in identifying patterns in complex medical data, leading to more accurate diagnoses (Bajwa et al., 2021). From analyzing radiology images to detecting anomalies in ECG patterns, AI is helping medical professionals make informed decisions. AI is also enabling a shift towards a more patient-centric model. AI is integral in creating a more responsive and effective healthcare ecosystem, from facilitating more precise diagnostics to predicting disease patterns and enabling personalized treatment plans to enhance patient care and engagement (Hall, 2023).

The advent of AI also marks a shift from "one-size-fits-all" treatment approaches to personalized medicine. Powered by AI, predictive analytics and genetic sequencing enable healthcare providers to tailor treatments to individual patients' genetic makeup, lifestyle, and environmental factors (Kizen, 2023). This customization level leads to more effective treatments and improved patient outcomes.

Moreover, AI's impact is more comprehensive than just clinical settings. It is helping shape healthier societies through proactive public health interventions and disease surveillance. AI's predictive capabilities are being used to model the spread of diseases, offering invaluable insights that can guide public health policies and interventions (Kizen, 2023).

AI is also revolutionizing patient care outside hospitals. The rise of AI-powered digital health solutions, such as telemedicine, wearable health tech, and AI chatbots, enhances access to healthcare services, promotes preventive healthcare, and enables better management of chronic conditions (Kizen, 2023).

Current Applications of AI in Healthcare

AI is reshaping numerous aspects of healthcare, from groundbreaking research and drug discovery to improved diagnostics and treatment methods, enhancing hospital management, and facilitating efficient post-treatment care (Kizen, 2023).

Diagnostics and Treatment

AI fundamentally transforms healthcare diagnostics and treatment strategies by leveraging machine learning algorithms for complex medical data analysis. These sophisticated algorithms are invaluable in scanning imaging scans, parsing laboratory results, and evaluating clinical notes. Their deployment enhances the precision and speed at which physicians can identify various diseases, including cancer, heart conditions, and neurological disorders (Roller, 2023).

AI's role is not confined to disease detection; it is also instrumental in formulating treatment plans. Predictive analytics, powered by AI, can forecast patient responses to treatments by considering their unique medical history, genetics, and lifestyle factors. This personalized therapy design approach can amplify treatment effectiveness and diminish undesired side effects (Roller, 2023).

Its capability to interpret complex medical data, including imaging scans, genetic profiles, and electronic health records, positions AI as a game-changer in healthcare. These algorithms surpass human abilities in analyzing imaging scans such as X-rays, CT scans, and MRIs, leading to high-accuracy detection of conditions like cancer, heart disease, or neurological disorders (Roller, 2023). Notably, Google's DeepMind developed an AI system that diagnoses eye diseases with the same accuracy as a top human expert. It highlights its potential to catch conditions that could escalate to blindness if untreated. Understanding this AI-driven transformation is vital for business and technology leaders as it represents a leap forward in healthcare efficiency and patient outcomes (Kizen, 2023). Moreover, AI applications in treatment plans are becoming increasingly sophisticated. By predicting patient responses to different treatments based on their unique medical history and genetic profile, healthcare providers can personalize treatment plans, improving outcomes and often reducing side effects (Kizen, 2023).

Research and Drug Discovery

AI's transformative influence extends to biomedical research and drug discovery, drastically streamlining these traditionally protracted and expensive processes. AI algorithms can process and analyze vast amounts of biomedical data, identifying potential molecular targets for drug development and predicting numerous compounds' therapeutic effects and adverse reactions (Kizen, 2023).

One application of AI lies in the swift identification of promising drug candidates. By analyzing the molecular structures of compounds, AI can predict potential therapeutic impacts and adverse reactions, significantly expediting the selection process for clinical trial candidates. AI also assists in modeling and simulating drug interactions, allowing researchers to anticipate possible side effects and drug-drug interactions early in the process (Paul et al., 2021).

AI's prowess is not limited to drug interactions; it is also invaluable in discovering patterns and correlations within vast data sets, tasks that would be prohibitively laborious or even impossible for humans. Companies like Insilico Medicine are harnessing this capability, using machine learning algorithms to design, synthesize, and validate novel drug candidates within mere days - a process traditionally spanning years (Paul et al., 2021).

Beyond laboratory data, AI can also analyze real-world data like health records and social health determinants. These insights facilitate the identification of new drug development targets and the prediction of potential drug side effects. Consequently, this reduces the time and cost involved in new drug introduction to the market, ensuring more effective treatments reach patients faster. As business and technology leaders, it is imperative to grasp AI's pivotal role in expediting drug discovery and enhancing patient treatment outcomes.

Hospital Management

Beyond its clinical applications, AI is also revolutionizing hospital operations and management. It is helping healthcare facilities become more efficient, reduce waste, and improve patient satisfaction. AI algorithms, for instance, can optimize surgery scheduling, considering surgeon availability, operating room utilization, and post-operative care resources. This enhances hospital efficiency and minimizes patient wait times, improving the patient experience (Mills, 2022).

Predictive analytics can also forecast patient flow in emergency departments. AI can predict patient influx by analyzing historical admission rates, seasonal disease patterns, and real-time local data, enabling hospitals to allocate their resources better (Kizen, 2023).

Furthermore, AI is proving instrumental in streamlining inventory management in hospitals. By predicting the demand for medical supplies based on factors like patient volume, seasonal trends, and disease outbreaks, AI can help hospitals maintain optimal inventory levels, reducing wastage and minimizing the risk of shortage.

AI in Rehabilitation

Integrating AI in rehabilitation opens new possibilities for patient recovery and well-being. AI-driven tools and devices can provide personalized therapy plans, monitor patient progress, and adapt strategies in real-time to maximize recovery (Kizen, 2023).

One exciting application is in physical therapy, where virtual reality combined with AI can create immersive, engaging therapeutic exercises. This approach can motivate patients, making therapy more enjoyable and effective.

Moreover, AI-powered assistive technologies can significantly enhance the quality of life for those with physical or cognitive disabilities. For example, AI-enabled communication devices can help those with speech impairments express themselves, promoting their social inclusion and independence ("AI Enhancing Human Experience in Healthcare," 2021).

AI in Medical Imaging

AI and deep learning's critical role in revolutionizing medical imaging – with applications spanning detection, diagnosis, delineation, and disease tracking – is becoming increasingly apparent. It has been instrumental in streamlining workflows for radiologists, automatizing labor-intensive tasks, and reducing the inherent variability in human interpretations. AI can process and interpret complex imaging data from different modalities, such as CT, MRI, and PET scans. This multimodal analysis can provide a comprehensive view of a patient's condition, supporting more informed clinical decision-making (Kizen, 2023).

However, procuring large datasets from a multitude of institutions, especially with regulations such as the Health Insurance Portability and Accountability Act (HIPAA) and a myriad of patient privacy and data ownership considerations, is a considerable challenge (Bajwa et al., 2021).

Federated learning, a promising alternative, can help overcome these issues by following a decentralized approach. Instead of data distribution, individual models are trained on local data in-house, and only these locally fine-tuned models are shared. This innovative approach safeguards patient data privacy, ensuring it remains well within the institutional boundaries during the deep learning process (Kwak & Bai, 2023).

Self-supervision in image-text pairs represents another breakthrough that's been catalytic in driving the advancement of robust, versatile vision-language AI models in biomedical niches like radiology. The vision is to harness auto-generated training signals – bypassing the need for manual labeling – so that models can independently pinpoint and map findings within images, correlating them with details from radiology reports (Kwak & Bai, 2023).

Radiologists routinely resort to comparative analysis with previous images for effective decision-making. Current AI frameworks in radiology, however, often fall short when tasked with synchronizing images with corresponding report data, chiefly due to limited access to preceding scans. A novel solution is Microsoft Research's BioViL-T – a self-supervised training framework that capitalizes on the temporal patterns inherent in biomedical datasets, thereby enhancing data efficiency (Kwak & Bai, 2023).

AI in Public Health Surveillance

AI holds substantial promise in transforming public health surveillance, aiding in predicting, monitoring, and managing health crises. It can efficiently analyze extensive and diverse data sets, including social media posts, weather patterns, and demographic information, to detect early signs of potential health issues. This predictive capacity can help healthcare authorities make proactive decisions, mitigating health risks and preventing crises before they occur (Jones, 2023).

Moreover, AI can play an instrumental role in monitoring environmental health hazards. By processing and interpreting complex environmental data, AI can detect pollution patterns, forecast potential risks, and facilitate interventions to protect public health. This proactive approach can enhance the overall health and wellness of communities (Baclic et al., 2020).

Furthermore, AI can be useful in understanding and addressing social determinants of health, such as income, education, and living conditions. These factors often significantly impact health outcomes, and AI's ability to analyze such complex relationships can help policymakers design effective health interventions and promote health equity.

AI in Health Education and Training

AI is revolutionizing healthcare education, offering innovative ways to train and develop healthcare professionals. Through virtual and augmented reality simulations, AI provides realistic, hands-on training experiences, allowing professionals to hone their skills in a safe environment (Lomis et al., 2021).

For instance, AI-powered virtual reality can simulate complex surgical procedures, giving medical students an immersive learning experience. These virtual procedures offer a risk-free platform for students to practice, learn from mistakes, and build confidence.

Moreover, AI can help design personalized learning pathways based on each learner's strengths and weaknesses. This individualized approach can improve learning outcomes, helping healthcare professionals better serve their patients (Lomis et al., 2021).

Personalized Medicine and AI

AI has the potential to usher in a new era of personalized medicine. By analyzing a patient's genetic makeup, lifestyle, and environment, AI can predict how they will respond to treatment more accurately, what potential health risks they might face, and what preventive measures would be most effective. This approach moves us from a "one-size-fits-all" model to a more personalized, patient-centered model of care (Schork, 2019).

In pursuing improved healthcare outcomes and patient experiences, the focus gradually shifts from a uniform approach to a personalized one. AI stands at the forefront of this transformative shift, demonstrating significant potential in personalized medicine. Through its ability to synthesize and analyze vast arrays of data, from genetic profiles to lifestyle factors, AI is redefining patient-centered care, paving the way toward a future where healthcare is as unique as the patients it serves (Payne, 2020).

In the past, medical professionals relied on generalized treatment protocols that, while effective for many, could not cater to individual nuances and specificities. However, AI is poised to change this through its capacity to analyze a patient's genetic makeup, lifestyle, and environmental influences in an integrated manner. The resultant insights allow for more accurate predictions of how a patient might respond to treatment, what potential health risks they might face, and what preventive measures would be most effective (Schork, 2019).

AI's role in personalized medicine extends beyond just tailored treatment plans. By leveraging AI, clinicians can better understand a disease's development and progression within the individual. For example, cancer, a complex disease that varies significantly among individuals, is an area where personalized medicine, supported by AI, is making substantial strides. AI algorithms can scrutinize tumor genetics, consider the patient's genetic information, and guide the selection of targeted therapies. This allows for treatments that are not only highly effective but also minimize the risk of adverse side effects (Kizen, 2023).

AI's prowess in predictive analytics also allows for proactive healthcare. By forecasting potential health risks based on a patient's unique health profile, AI helps physicians anticipate medical conditions and intervene early (Kizen, 2023). Such predictive capabilities extend to chronic diseases such as heart disease and diabetes, allowing for lifestyle modifications and preventive measures to be implemented before the disease manifests.

Additionally, AI plays a significant role in drug compatibility and personalized drug dosing. Different patients metabolize drugs at different rates, and AI can help determine the most effective dosage for a specific patient, reducing the risk of under-dosing or overdose.

Another exciting avenue for AI in personalized medicine is digital health apps and wearable tech. AI-powered apps can provide users with health and wellness recommendations based on personal data, ranging from diet and exercise habits to sleep patterns and stress levels. Similarly, wearable devices can continuously monitor vital signs and physiological parameters, enabling real-time health management tailored to the individual's needs (Kizen, 2023).

This shift from a "one-size-fits-all' model to a more personalized, patient-centered model of care is reshaping healthcare. Integrating AI in personalized medicine leads to better health outcomes and improved patient experiences and aligns with the larger goal of making healthcare more proactive rather than reactive (Kizen, 2023).

In this new paradigm of care, every patient's health journey is unique, with treatments, preventive strategies, and health advice tailored to their needs and circumstances. With its ability to analyze and learn from vast amounts of diverse data, AI is the powerful engine driving this transformative shift toward truly personalized medicine (Kizen, 2023).

AI in Mental Health

Mental health, a critical component of our overall well-being, often remains under-addressed due to various factors ranging from stigma to lack of access to quality mental health care. However, the emergence of AI in this sphere promises a revolution. From AI-powered chatbots providing constant psychological support to machine learning algorithms spotting early signs of mental health disorders, AI is reshaping mental health care, fostering its accessibility and timely intervention (Philips, 2021).

AI-powered chatbots are proving to be a crucial asset in providing psychological support. These virtual assistants, built on natural language processing and machine learning algorithms, can converse with users in a human-like manner, providing emotional support, guiding mindfulness practices, and even teaching cognitive behavioral techniques. Their round-the-clock availability makes mental health support accessible anytime, anywhere, and their anonymity encourages individuals hesitant to seek help due to societal stigma (Muller, 2021).

Moreover, these chatbots can identify signs of severe distress or crisis during conversations, alerting human therapists who can intervene. This blend of AI and human oversight can be particularly helpful in preventing problems and managing mental health emergencies.

Machine learning algorithms are also making strides in the early detection of mental health conditions. These algorithms can analyze cues from speech patterns, text analysis, or social media behavior to detect signs of mental health issues such as depression, anxiety, or bipolar disorder. Subtle changes in speech, tone, word choice, or online behavior shifts can be early indications of a developing mental health issue. AI's ability to detect these nuances can enable timely intervention, preventing the exacerbation of these conditions (Bateman, 2021).

In addition to detection, machine learning, and AI can assist in predicting mental health crises or assessing suicide risk. By analyzing factors like previous mental health history, social determinants, and personal narratives, AI can help clinicians anticipate problems and take preemptive measures.

AI is also helping bridge the treatment gap in mental health. Teletherapy platforms, many of them powered by AI, are increasing access to mental health services, especially in remote or underserved regions. These platforms facilitate online counseling sessions, group therapies, and self-help programs, making mental health care more accessible and convenient (Bateman, 2021).

Moreover, AI's role in personalizing mental health treatment cannot be overstated. AI algorithms can analyze individual patient data to determine the most effective therapeutic approach, considering the person's unique experiences, symptoms, and response patterns. Such tailored treatment plans can lead to more successful outcomes and better management of mental health disorders.

As we leverage AI in mental health care, we must also tread with caution. Issues of privacy and consent are critical, given the sensitive nature of mental health data. Establishing robust data privacy practices and transparent AI algorithms is crucial to ensuring ethical AI applications in mental health.

In summary, AI's role in mental health care is transformative. It can make mental health care more accessible, personalized, and effective. By leveraging AI, we are moving towards a future where mental health care is timely, inclusive, and proactive, contributing to healthier, more resilient communities.

Future Potential of AI in Healthcare

As AI continues to evolve and mature, its potential to transform the healthcare sector grows exponentially. From aiding in genomics and precision medicine to addressing global health challenges, enhancing prosthetics, and supporting aging populations, AI promises profound advancements in understanding and managing health.

Genomics and Precision Medicine

The field of genomics, a branch of molecular biology focused on studying an organism's entire set of genes, is ripe for AI intervention (Kizen, 2023). As we gain a better understanding of the human genome, AI has a crucial role in interpreting vast genetic data, thereby facilitating the development of personalized therapies (Johnson et al., 2021).

AI can help decipher patterns within genetic sequences, identify disease-linked mutations, and predict how different genes may interact or respond to certain drugs. This can facilitate precision medicine, where treatments are designed based on an individual's genetic makeup, providing a more targeted and practical approach to disease management (Kizen, 2023).

Global Health

AI also holds significant promise for global health, particularly in tackling pressing health challenges like disease outbreaks or healthcare inequities. AI algorithms can predict disease outbreaks by analyzing weather patterns, population movement, and social media posts. Early warning can enable proactive response measures, limiting the spread and impact of diseases ("Artificial Intelligence in Global Health," 2022).

In low-resource settings, AI can help improve healthcare delivery by optimizing scarce resources, facilitating remote consultations, and offering predictive insights to anticipate health crises. Addressing these areas, AI can be crucial in reducing health disparities and achieving health equity globally.

Advanced Prosthetics

The domain of prosthetics is another area poised for an AI-powered transformation. Prosthetic limbs that respond to neural signals made possible through AI integration, can significantly improve the quality of life for individuals with limb loss. These "smart" prosthetics can replicate various natural movements, enabling users to perform everyday activities efficiently. Additionally, they hold the potential to provide sensory feedback, such as touch or temperature, further enhancing their functionality (Bryant, 2019).

Aging and Home Care

With aging populations in many countries, AI can significantly ensure older adults' independence and quality of life. AI-powered home monitoring systems can track vital signs, detect falls, or alert caregivers of anomalies in the resident's routine. Robotic caregivers can aid with daily tasks, promote medication adherence, and offer social interaction, addressing some of the challenges associated with aging (Corbyn, 2021).

Moreover, integrating AI with the Internet of Things (IoT) can provide real-time health insights and foster better communication with caregivers (Meyer,

2021). Advanced mobile applications paired with IoT devices can monitor health parameters, provide reminders for medication, and offer virtual consultations, facilitating continued care and interaction with healthcare providers (Meyer, 2021).

Despite this vast potential, the successful implementation of AI in healthcare will require addressing several challenges. Data privacy and security are paramount, given the sensitive nature of health data. Algorithmic bias, where AI systems may reflect or amplify societal biases, can lead to disparities in healthcare outcomes. Moreover, there is a pressing need for appropriate regulatory oversight to ensure AI's safe and ethical use in healthcare.

Robotics in Surgery

AI plays an increasingly significant role in surgery, enhancing precision, and reducing recovery time. Robotic surgical systems, guided by AI and computer vision, can perform intricate procedures with remarkable accuracy (Kizen, 2023). These systems can navigate tight spaces, make minute incisions, and reduce the risk of complications, leading to safer surgeries and faster patient recovery (Nathan, 2023).

AI can also assist in surgical planning. By analyzing medical images, AI can help surgeons visualize the procedure, identify potential challenges, and plan the best course of action (Nathan, 2023).

Additionally, AI can help monitor patient vitals during surgery, alerting the surgical team to any significant changes. This real-time monitoring can help ensure patient safety and optimize surgical outcomes (Nathan, 2023).

Patient Engagement and Experience

AI is reshaping patient engagement and experience, offering personalized, convenient, and efficient care. AI chatbots, for instance, can answer patient queries, schedule appointments, provide health advice, and deliver round-the-clock customer service (Bajwa et al., 2021).

Wearable technology, another AI-driven innovation, allows continuous monitoring of health parameters. These devices can alert patients and their healthcare providers to potential health issues, promoting proactive care ("Artificial Intelligence in Global Health," 2022).

Furthermore, AI can personalize patient communication based on individual preferences and needs. This personalized approach can improve patient satisfaction, engagement, and overall healthcare experience ("Artificial Intelligence in Global Health," 2022).

Case Studies in AI Applications in Healthcare

There are numerous examples of the transformative potential of AI in healthcare. For example, Google's DeepMind developed an AI system to diagnose eye diseases by analyzing retinal images, matching, or exceeding human expert performance (Ram, 2018).

Another compelling case is Zebra Medical Vision, an AI start-up that uses machine learning to read medical imaging data. Their algorithms can detect lung cancer, cardiovascular events, and liver disease with remarkable accuracy.

Not all applications, though, were successful. For instance, IBM Watson for Oncology gave "unsafe and incorrect" cancer assessments (Chen, 2018). It was eventually "sold for parts (O'Leary, 2022)."

Conclusion

Artificial Intelligence is reshaping healthcare in profound ways. From diagnostics and treatment to patient engagement and experience, AI is making healthcare more personalized, efficient, and accessible. AI's ability to analyze and interpret vast amounts of data enables proactive healthcare, predicting health risks and facilitating early interventions. Despite challenges related to data privacy and algorithmic bias, the potential of AI in healthcare is immense. As we continue to harness this potential, we move towards a future where healthcare is increasingly proactive, personalized, and patient-centered.

Chapter 12

AI IN MARKETING, SALES, LEAD GENERATION, AND CUSTOMER ACQUISITION

The intersection of artificial intelligence (AI) and business practices is reshaping marketing, sales, lead generation, and customer acquisition. AI equips marketers and salespeople with the tools to deeply understand and engage customers, fostering an environment for effective business-client relations (Davenport et al., 2021).

AI introduces an era of personalization at scale. Customer behavior data, dissected by AI algorithms, uncover each individual's unique preferences. These insights shape tailored marketing messages, sparking engagement and boosting conversion rates.

Customer service sees its evolution expedited with AI. AI-powered chatbots and virtual assistants are at the frontline of customer service, answering routine inquiries round-the-clock. This liberates human agents to address complex issues. Simultaneously, machine learning algorithms examine customer interaction data, pinpointing areas ripe for improvement (Payani, 2023).

Campaign optimization is another area where AI shines. When fed into AI algorithms, performance data from marketing campaigns help adjust campaign parameters in real time. This dynamic adjustment maximizes return on investment, ensuring more efficient movements and prudent use of marketing budgets.

In a world where data abounds, utilizing it efficiently is a competitive edge. With its powerful toolkit - machine learning algorithms, natural language processing, and predictive analytics - AI affords businesses more profound insights into customer behavior and preferences. This understanding transforms engagement strategies and makes winning over customers a precise art.

This AI-led transformation is more than a mere efficiency upgrade – it redefines the domains of marketing, sales, lead generation, and customer acquisition. It transitions businesses from fragmented operations to an integrated, customer-centric model capable of dynamically adapting to ever-changing customer needs and desires. As business and technology leaders, comprehending and leveraging this AI-induced transformation is the key to navigating the future of business (Payani, 2023).

However, harnessing the power of AI in these areas is challenging. It requires a deep understanding of the technology and the evolving customers' needs and wants. As such, marketers, salespeople, and business leaders must stay ahead of the curve, exploring and investing in AI capabilities while ensuring ethical and responsible use of this powerful technology.

In the following sections, we explore how AI is used in marketing, sales, lead generation, and customer acquisition, explore its potential benefits, and offer guidance on navigating the complexities of implementing AI in these areas. From AI-powered customer segmentation and personalized marketing to predictive sales analytics and intelligent lead scoring, we will examine how AI is changing the game, offering exciting possibilities for businesses to connect with customers in novel and impactful ways.

AI in Marketing

The role of AI in marketing is expansive and transformative, spanning from predicting consumer behavior, optimizing campaigns, and aiding in content creation to powering personalization at scale. It provides the necessary tools to navigate the ever-growing complexity of marketing, and more importantly, it helps forge more robust, more meaningful relationships with consumers.

Consumer Behavior Prediction

Understanding what consumers want, how they behave, and what they will do next is crucial in marketing. AI can analyze vast amounts of data—

purchases, web browsing activity, social media interactions, and more—to identify patterns and predict consumer behavior. These AI-driven insights can inform marketing strategy, helping businesses to target their audiences better, tailor their messaging, and anticipate future trends (Starita, 2022).

Machine learning models can predict what a customer is likely to buy when they are likely to buy it, and what marketing actions will most likely prompt that purchase. They can also help marketers identify high-value customers, understand customer churn, and detect changes in customer sentiment (Gkikas & Theodoridis, 2021). This level of insight empowers businesses to be proactive rather than reactive, staying one step ahead of customer expectations and market trends. For example, AutomotiveMastermind, a subsidiary of the S&P Global transportation division, employs AI and predictive analytics to analyze consumer buying patterns to anticipate future purchases. This technology enables dealerships to use their first-party data, CRM tools, and third-party data from various sources to accurately pinpoint consumers who are likely to buy a car soon, and even predict the make and model. This precise identification supports local dealerships in deploying targeted campaigns, both physical and digital, to efficiently influence the purchasing decision.

Campaign Optimization

Marketing campaigns involve many parameters, including the target audience, timing, messaging, channel selection, and budget allocation. AI algorithms can manage and optimize these elements in real time, adjusting parameters based on performance to maximize the return on investment (Davenport et al., 2021).

For example, programmatic advertising platforms use AI to automatically buy and optimize digital ad placements, making real-time decisions about where to display ads to achieve the highest conversion rates. Similarly, AI can optimize email marketing campaigns, determining the best time to send emails to each recipient and tailoring the content to their interests (Gupta, 2022).

Content Creation

Creating engaging content is a crucial aspect of marketing, and AI can assist in this process, making it faster, more efficient, and more personalized. AI can generate basic content such as news articles, blog posts, and social media updates, saving marketers valuable time and effort. Solutions like WriteSonic enable marketers, bloggers, and SEO specialists to generate content in bulk with simple keywords and topics of focus (Hernandez, 2023).

More advanced AI applications are even venturing into creating video content. AI-powered platforms can transform scripts into animated videos, while other tools use AI to edit and optimize video content based on audience engagement data. While AI cannot replace human creativity, it can handle the more routine aspects of content creation, freeing marketers to focus on strategic and creative tasks. Synthesia, for instance, is a digital media platform that lets users create AI-generated videos (Singh, 2023). The London-based company raised $90 million in a recent funding round, backed by U.S. chipmaker Nvidia. Synthesia's software allows people to make their own digital avatars to deliver corporate presentations, training videos, or even compliments to colleagues in more than 120 different languages. "Its ultimate aim is to eliminate cameras, microphones, actors, lengthy edits, and other costs from the professional video production process (Singh, 2023)."

AI in Sales and Lead Generation

AI is increasingly playing a crucial role in sales and lead generation. By harnessing the power of AI, businesses can enhance efficiency, predict outcomes more accurately, personalize outreach, and significantly increase the likelihood of conversion. Here, we explore these applications, illustrating how AI can revolutionize sales and lead generation processes.

Sales Forecasting

Sales forecasting is critical to strategic planning, helping businesses anticipate sales revenue, manage resources effectively, and set realistic goals. AI elevates this process by analyzing vast amounts of historical sales data, identifying patterns and trends, and predicting future sales with remarkable accuracy (Sinha et al., 2023).

Advanced machine learning models can consider many factors – including seasonal trends, economic indicators, market conditions, and changes in customer behavior or product preferences. By offering a nuanced and data-driven view of future sales, AI-enabled forecasting allows businesses to be better prepared, make informed decisions, encourage proactive outreach to consumers based on product and service consumption, and ultimately stay competitive (Sinha et al., 2023).

Lead Scoring

Every prospect has a different value to the business, and knowing which leads are most likely to convert can significantly enhance the effectiveness of sales efforts. AI is pivotal here, enabling enterprises to score or rank potential customers based on their conversion likelihood.

Machine learning algorithms can analyze various data points about each lead, with comparison to their cohorts, including demographic information, online behavior, engagement with marketing activities, and more. By scoring leaders this way, AI helps sales teams prioritize their efforts, focusing on the most promising prospects and improving conversion rates and sales efficiency (Wisneski, 2021).

Personalized Outreach

AI is not only about streamlining processes and predicting outcomes, but it is also about enhancing human relationships – a fundamental aspect of sales. AI can assist salespeople in personalizing their outreach, equipping them with deep insights into individual prospects, and suggesting effective communication strategies (McBeth, 2023).

AI can help sales teams understand each prospect's unique needs, preferences, and pain points, drawing on data from CRM systems, social media, and other sources. AI tools can also recommend the best time to reach a prospect, the most effective communication channel, and even suggest personalized messaging. This level of personalization can significantly enhance the effectiveness of sales outreach, fostering stronger relationships with prospects and increasing the likelihood of conversion.

In summary, AI is significantly transforming sales and lead generation, bringing unprecedented levels of efficiency, accuracy, and personalization. By leveraging these AI capabilities, businesses can enhance their sales processes, boost conversion rates, and drive revenue growth.

AI in Customer Acquisition

Customer acquisition is a critical business process, and with the advent of AI, this process is becoming increasingly data-driven, targeted, and effective. AI can improve targeting, bolster engagement, and optimize conversion rates, significantly enhancing a business's capacity to attract and retain customers.

Targeting

Accurate targeting is foundational to any customer acquisition strategy. It allows businesses to direct their marketing efforts toward those most likely interested in their product or service. AI can elevate this process by helping companies to identify their ideal customer profile and find potential customers who match it.

Machine learning algorithms can analyze vast amounts of data, pulling from internal data (e.g., customer demographics and purchase history) and external data (e.g., market trends and social media behavior) to create an accurate and dynamic picture of the ideal customer. Businesses can use this AI-generated customer profile to target potential customers more effectively through personalized marketing campaigns, search engine optimization, or social media advertising (Frąckiewicz, 2023).

Engagement

Engaging potential customers is equally essential in the customer acquisition process. Engagement builds awareness, fosters relationships, and ultimately nudges potential customers toward conversion. AI can assist in driving engagement by creating personalized content and interactive experiences.

AI-powered content creation tools can generate personalized content tailored to each potential customer, reflecting their unique interests, needs, and behaviors. Additionally, AI-powered chatbots can provide interactive experiences, respond to real-time customer queries, guide them through the purchasing process, and provide personalized recommendations (Starita, 2022). This high level of personalization and interaction can significantly boost engagement, strengthening relationships with potential customers and making them more likely to convert.

Conversion

Finally, AI can optimize the conversion process itself. Conversion is the critical moment when a potential customer decides to purchase. AI can enhance this process by identifying and addressing barriers to conversion and suggesting ways to improve the overall customer experience (McBeth, 2023).

By analyzing customer behavior data, AI can identify points in the customer journey where potential customers are most likely to drop off. Businesses can then use these insights to address issues and remove obstacles by simplifying the checkout process, offering personalized incentives, or enhancing customer support. Moreover, AI can provide continuous optimization, using real-time data to adjust strategies and improve the customer experience dynamically.

Enhancing Customer Satisfaction, Loyalty, and Engagement

In an increasingly competitive business environment, customer satisfaction, loyalty, and engagement have become key differentiators. AI-driven personalization offers companies the ability to elevate the customer experience in a meaningful and impactful way (Frąckiewicz, 2023). In the future, digital experiences through web and mobile could become completely personalized through their look and feel, offering a unique digital experience to every user. Through tailored experiences, anticipation of customer needs, seamless cross-channel experiences, real-time interaction, dynamic pricing, emotional connections, predictive customer service, and continuous improvement, AI-driven personalization paves the way for greater customer satisfaction, loyalty, and engagement.

Tailored Experiences and Anticipating Customer Needs

AI technology provides businesses with the tools to create personalized experiences that cater to customers' unique preferences and needs. This is achieved by analyzing vast amounts of customer data, including purchase information, browsing behavior, and demographic data. Furthermore, AI's predictive analytics capabilities can anticipate customer needs and preferences based on past behaviors and patterns, proactively offering relevant products, services, or content (Frąckiewicz, 2023). The ability to deliver tailored experiences and proactively address customer needs enhances satisfaction and fosters customer loyalty.

Seamless Cross-Channel Experiences and Real-Time Interaction

AI ensures consistency in customer experience across various channels and touchpoints. Whether through websites, mobile apps, social media platforms, or offline interactions, AI aids in delivering personalized recommendations and a cohesive brand experience. Additionally, AI-powered chatbots and virtual assistants enhance real-time customer interaction, providing instant support and personalization, effectively reducing response times and meeting customer needs promptly ("Omni-channel AI," 2023).

Dynamic Pricing, Emotional Connection, and Predictive Customer Service

Leveraging AI, businesses can optimize their pricing strategies and offers in real-time, based on market data, demand patterns, and individual customer profiles (Dilmegani, 2023). At the same time, AI's ability to analyze customer sentiment and emotional cues aids businesses in establishing a deeper emotional connection with their customers, leading to increased satisfaction and loyalty (Zaki et al., 2021). Predictive customer service, another aspect of AI's capabilities, allows businesses to foresee customer needs and preemptively address potential issues, enhancing customer service efficiency and fostering long-term loyalty.

Continuous Improvement and the Balance of Personalization and Privacy

AI's strength lies in its capacity for continuous learning and adaptation based on customer feedback and preferences. As businesses strive to offer increasing levels of personalization, striking a balance with privacy becomes critical. Clear communication about data usage, robust data protection measures, and obtaining appropriate consent are essential to maintain customer trust.

The potential of AI technologies to provide unprecedented advantages in marketing, sales, and customer acquisition has been established. However, the realization of these benefits hinges on the effective implementation of these technologies.

Best Practices for Implementing AI-Powered Technologies

To unlock the full potential of personalization and create experiences that truly resonate with customers, businesses need to implement AI-powered technologies effectively. These technologies include recommendation engines, chatbots, and customer analytics. In the following sections, we cover best practices for leveraging each of these tools to optimize customer experiences.

AI-Powered Recommendation Engines

The effectiveness of recommendation engines relies heavily on the quality and relevance of the data used to train the algorithms. Additionally, the choice of personalization algorithms, as well as transparency and control provided to customers about data usage and privacy, play pivotal roles in the engine's success.

AI-Powered Chatbots

Natural Language Processing (NLP): Implement chatbots with robust NLP capabilities to accurately understand and respond to customer inquiries. Continuously train chatbots with real-time data to improve their language understanding and conversational abilities.

Seamless Handover to Human Agents: Design chatbots to seamlessly transfer conversations to human agents when necessary. This ensures a smooth transition and provides customers with a seamless, personalized experience across channels.

Constant Learning and Improvement: Enable chatbots to learn from each customer interaction to improve their responses and handle a broader range of inquiries over time. Regularly analyze chatbot performance, identify areas for improvement, and implement necessary updates.

Customer Analytics

Data Integration: Integrate customer data from multiple sources, such as CRM systems, social media platforms, and transactional data, to gain a holistic view of each customer. Consolidate and centralize data to enable comprehensive customer analytics.

Advanced Segmentation: Utilize AI-driven customer segmentation techniques to create targeted customer segments based on behavioral, demographic, and psychographic attributes. This enables personalized marketing campaigns and tailored customer experiences.

Predictive Analytics: Capitalize on predictive analytics to foretell customer behavior, pinpoint emerging trends, and refine marketing and sales tactics. Employ machine learning algorithms to envisage customer preferences and anticipate future actions, ultimately informing business strategy and enhancing customer engagement.

Real-time Insights: Integrate real-time customer analytics to gather and scrutinize customer data instantaneously. This swift data capture and analysis enable businesses to promptly cater to customer needs and tailor experiences, thus driving customer satisfaction and loyalty. By staying responsive and adaptive, companies can maintain a solid competitive edge in today's dynamic market.

Ethical Considerations

Transparency and Consent: Communicate clearly with customers about how their data will be used to enhance their experiences. Obtain explicit consent and provide opt-out options for customers who do not wish to participate.

Data Privacy and Security: Implement robust data privacy and security measures to protect customer data from unauthorized access or misuse. Comply with relevant regulations and industry standards to safely handle customer information.

Continuous monitoring, analysis, and optimization are essential for all AI-powered tools. Regularly assess the performance and effectiveness of recommendation engines, chatbots, and customer analytics to identify areas for improvement and make necessary adjustments.

By implementing these best practices, businesses can leverage AI-powered recommendation engines, chatbots, and customer analytics to optimize customer experiences. Delivering personalized recommendations, seamless conversations, and data-driven insights enables enterprises to engage customers effectively, build loyalty, and drive long-term success in a competitive market.

Future of AI in Marketing, Sales, and Customer Acquisition

Looking into the future, it is evident that AI will continue to revolutionize these fields. Businesses wishing to stay competitive must anticipate and prepare for upcoming trends, including integrating online and offline customer experiences, predictive personalization, and the increasing importance of AI ethics and regulation.

Integrating Online and Offline Experiences

One of the critical future directions of AI in these fields is integrating online and offline customer experiences, ultimately creating a unified and seamless customer journey. Currently, many businesses treat their online and offline channels as separate entities. However, customers increasingly expect a seamless experience as they interact with companies across various touchpoints.

AI technologies can help bridge this gap by providing a holistic view of the customer's journey. For example, AI can use data from online interactions (like website visits or social media engagements) and offline interactions (like in-store purchases or call center inquiries) to create a comprehensive customer profile. This profile can personalize customer interactions across all touchpoints, enhancing customer satisfaction and loyalty.

Predictive Personalization

As AI algorithms evolve and improve, we will shift towards predictive personalization. AI will respond to customers' needs and preferences and anticipate their future.

With the ability to analyze vast amounts of data in real time and learn from patterns, AI can predict what customers want before they know it themselves. This can manifest in various ways - anticipating the next product a customer might want to buy, foreseeing when they need customer service, or even proactively addressing concerns before they arise. This level of predictive personalization will lead to even more individualized customer experiences and foster deeper customer relationships.

AI Ethics and Regulation

With the increasingly ubiquitous use of AI in engaging with customers, ethical considerations and regulatory compliance have become even more critical. Businesses must navigate these issues carefully, considering data privacy, transparency, and fairness.

Data privacy is paramount as businesses handle vast amounts of personal data. Companies must ensure they are using AI to respect privacy and protect data, adhering to regulations like GDPR and CCPA. Transparency is another essential aspect, especially in how AI algorithms make decisions. Customers need to understand why they receive specific recommendations or why certain data is being collected. Finally, ensuring fairness and avoiding bias in AI decision-making is crucial, as algorithms that unintentionally discriminate can harm both customers and the business's reputation.

AI holds vast potential for the future of marketing, sales, and customer acquisition. However, businesses must approach AI adoption thoughtfully and strategically to realize this potential. The future belongs to those businesses that understand and embrace the power of AI, navigating its challenges and leveraging its possibilities to create compelling and personalized customer experiences. As they do so, they will be well-positioned to succeed in an increasingly competitive business environment.

Chapter 13

AI IN FINANCIAL SERVICES

The financial services sector has been a pioneer in incorporating Artificial Intelligence (AI). "AI's ability to process vast amounts of data, identify patterns, and make predictions has made it invaluable to augment the value-add of traditional financial services (Savi et al., 2023)," whether for business-to-business (B2B) or business-to-consumer (B2C) models, from risk assessment frameworks to customer service interactions.

Financial institutions increasingly leverage AI to streamline operations, enhance customer experiences, and drive innovation. However, AI's adoption is not uniform across the sector, with some areas witnessing more rapid adoption than others. As AI technologies evolve, their impact on the financial industry is expected to grow, ushering in a new era of innovation and transformation.

Impact of AI on Various Financial Services

AI in Banking

The banking sector has been at the forefront of AI adoption, with institutions leveraging AI to streamline operations, enhance customer experiences, and drive innovation. AI is used in credit decision-making, where it helps banks assess customers' creditworthiness more accurately and efficiently. AI also helps enhance customer service, with chatbots and virtual assistants providing round-the-clock support and personalized services.

AI is also playing a crucial role in the fight against financial crime. A prime example is Google Cloud's AI-powered anti-money laundering (AML) offering. The tool uses AI to analyze transaction data, identify suspicious patterns, and flag potential money laundering cases. This helps banks comply with regulatory requirements and protects them from financial and reputational damage.

AI is also enhancing the banking experience by augmenting the traditional banking model. For example, AI can analyze user activities and data from other non-banking apps to offer customized financial advice. Banks like The Development Bank of Singapore (DBS) and the Royal Bank of Canada (RBC) have already embraced AI-based tools to that effect. For instance, RBC has developed a NOMI platform that helps customers automate savings and manage their monthly budgets effectively. "The platform has 1.5 million active users, 53% of whom consider it a game-changer for their finances (Baheti, 2022)." One of NOMI's flagship features, Find&Save, has helped customers save about 1.9 billion dollars by rounding up expenses and automatically transferring small change to savings accounts.

AI in Credit Worthiness

AI transforms allocating credit and risk, resulting in fairer and more inclusive systems. According to Forbes, 70% of financial firms use machine learning to predict cash flow events and adjust credit scores. AI enables better credit systems by developing a system where lenders can more accurately determine a borrower's risk with the aid of AI, regardless of social-demographic conditions. This can be accomplished by looking at real-time indicators not considered in a typical credit score, such as whether the borrower spends their money on necessities or luxuries, income level, employment opportunities, and earning potential. "AI may also assist lenders in identifying less visible risk characteristics, such as whether a borrower exploits their available credit (Baheti, 2022)."

Leading financial technology companies, or FinTech, like JP Morgan, have clarified that the future of customer-centric financial services lies in crunching vast amounts of data drawn from varied sources—often non-traditional. "J.P. Morgan has recently summarized critical research in machine learning, big data, and artificial intelligence, highlighting exciting trends that impact the financial community (Baheti, 2022)." AI might eventually be able to completely replace current mathematical credit scoring systems that get much criticism for being outdated—primarily because of their standardization and lack of sensitivity to individual disparities and nuances.

AI in Fraud Detection and Risk Management

AI-based systems are helping consumers minimize the risk and save money from fraudulent activities. AI applications range from recognizing abnormal transactions to identifying suspicious and potentially fraudulent activities. AI can also help financial institutions manage risk more effectively. By analyzing vast amounts of data, AI can identify patterns and trends to help institutions make more informed decisions about risk. For example, AI can help institutions identify risky investments or loans, helping them avoid potential losses.

However, using AI in fraud detection and risk management is challenging. AI models must be trained on vast amounts of data, which can be costly and time-consuming. Additionally, AI models must be explainable to ensure fairness and transparency. This is especially important when AI is used to make decisions that affect individuals, such as determining whether a transaction is fraudulent or a loan application is rejected.

Risk modeling and fraud detection are two areas where AI has shown significant promise in financial services. Traditional AI systems have been used for these tasks for years, but generative AI brings new capabilities. For instance, generative AI can handle more complex and nuanced scenarios, making it a valuable tool for identifying and mitigating risks.

However, using AI in risk modeling and fraud detection raises essential data privacy and security questions. Financial institutions must ensure that their use of AI complies with all relevant regulations and that customer data is always protected. Despite these challenges, the potential benefits of AI in risk modeling and fraud detection are significant, and we will likely see more financial institutions adopting these technologies in the future.

AI in Wealth Management

The wealth management sector is also witnessing a surge in AI adoption. The arrival of AI and ChatGPT in wealth management is a game-changer, enabling wealth managers to offer more personalized and efficient services. AI can help wealth managers provide more customized services to their clients. By analyzing client data, AI can help wealth managers better understand their client's needs and preferences, enabling them to provide more tailored advice and services. For example, AI can help wealth managers identify investment opportunities that align with their client's risk tolerance and investment goals.

AI's role in wealth management is more comprehensive than providing investment advice. It is also being used to streamline operations and enhance customer experiences. For instance, AI-powered chatbots are used to provide round-the-clock customer support, while AI algorithms are used to automate portfolio management and rebalancing.

However, using AI in investment and wealth management has its challenges. AI models must be trained on vast amounts of data, which can be costly and time-consuming. Additionally, AI models must be explainable to ensure fairness and transparency. This is especially important when AI makes decisions affecting individuals, such as determining whether an investment suits a client.

AI in Accounting and Finance

The accounting and finance sector is also benefiting from AI adoption. AI automates routine tasks, analyzes financial data, and provides actionable insights. This enhances efficiency and enables accountants and finance professionals to focus on more strategic tasks. For example, AI can now categorize and classify transactions, reconcile receipts, manage expense reports, sort transactions, and capture anomalies with data recording and reporting.

AI's impact on accounting and finance is not limited to automation. It is also helping firms make more informed decisions. For instance, AI algorithms can analyze vast amounts of financial data, identify trends, and provide predictive insights. AI can prepare financial and income statements for finance teams, generate boilerplate language for SEC reporting, or prepare remarks for analysts' calls.

Trade Settlement Process Automation

AI and ML are transforming trade settlements by enhancing operational efficiencies and reducing costs. Before the introduction of ML/AI in finance, teams at financial institutions would need to process the trade failure, identify the reason, and resolve the issues. "AI technologies can analyze vast amounts of historical trading data to identify failed trades, a process that is significantly faster and more efficient than manual analysis by back-office personnel (Patel, 2018)." Once the failed trades are identified, AI can analyze the reasons for their rejection and spot anomalies, gaps, or missing elements in the settlement process. AI can also predict the likelihood of future trade failures and propose solutions through pattern recognition analysis, improving its predictive capability over time. This leads to a "more efficient settlement process, fewer exception processing situations, and lower operational costs (Patel, 2018)." AI can also act as an added layer of security in catching failed trades, enhancing back-office operational efficiencies. According to a report from Goldman Sachs, machine learning and AI could enable $34 billion to $43 billion in annual savings and revenue opportunities within the financial sector by 2025 (Patel, 2018).

AI in Market and Consumer Sentiments

Numerous hedge funds and quantitative analysts have crafted strategies for trading in the financial markets by utilizing indicators derived from news sources as well as social media sentiment, confidence levels, and story-share counts. "However, traditional event-driven investment strategies and surveillance methodologies rely on mining for known behavior and patterns (Falk, 2023)." Generative AI has the potential to surface new themes and associated sentiments without direction. For instance, LLMs can identify new trends in consumer behavior from social media content by clustering posts with similar meanings and assigning the clusters an aggregate measure of sentiment. "Similarly, negative sentiment associated with specific content, such as a new advertising campaign, can quickly be identified and summarized. Investors and enterprises can then respond promptly to this information (Falk, 2023)."

AI in Customer Service

Customer service is another area where AI is making a significant impact in the financial services sector. Traditional AI systems have been used to handle straightforward customer queries, but generative AI can handle more complex and bespoke requests. This can help improve customer satisfaction and reduce the cost of customer service.

Generative AI can provide more accurate and responsive answers to customer queries, making it a valuable tool for improving customer service. It can also help customer service representatives by providing them with better information, enabling them to respond to customer queries more effectively and efficiently.

However, using AI in customer service raises data privacy and security questions. Financial institutions must ensure that their use of AI complies with all relevant regulations and that customer data is always protected. Despite these challenges, the potential benefits of AI in customer service are significant, and we will likely see more financial institutions adopting these technologies in the future.

AI and Job Impact on Financial Services

The integration of AI into financial services is significantly impacting jobs in the sector. On the one hand, AI is automating routine tasks, reducing the need for human intervention. On the other hand, it is creating new roles and opportunities for individuals with AI and data science skills.

The impact of AI on jobs in financial services is not uniform. While some roles are being automated, others are being enhanced. For instance, AI is automating tasks such as data entry and basic customer service, but it is enhancing roles that require complex decision-making and strategic thinking. This is leading to a shift in the skills needed for the financial sector, with a growing demand for individuals with skills in AI and data science.

Integrating AI into financial services also has a broader impact on the job market. It is driving the creation of new roles and opportunities in areas such as third-party plug-ins and services, AI development, data science, and AI ethics. At the same time, it is driving the transformation of existing functions, with individuals needing to upskill and reskill to stay relevant in the AI-driven job market.

Ethical Considerations and Responsible AI

The integration of AI into financial services is not without ethical considerations. These include issues related to privacy, fairness, and transparency. For instance, AI algorithms used in credit decisions must be fair and transparent, and they should not discriminate against certain groups of customers.

The concept of responsible AI is gaining traction in the financial sector. This involves developing and using AI in an ethical, transparent, and accountable way. For instance, financial institutions are implementing measures to ensure that their AI algorithms are fair and transparent, and they are taking steps to mitigate the risk of AI bias.

Despite the challenges, the financial sector is progressing in addressing the ethical considerations associated with AI. This is driven by regulatory pressure, customer expectations, and the recognition that responsible AI is good for business. As the sector continues to navigate the ethical complexities of AI, it is expected to play a leading role in shaping the future of responsible AI.

Future of AI in Financial Services

Democratization of Financial Services through AI

AI is playing a pivotal role in the democratization of financial services. The FinTech industry increasingly focuses on providing services "for the unbanked or underbanked, including people with disabilities, minorities, and marginalized groups (McKendrick & Margaris, 2023)." Many financial services that are taken for granted by some are inaccessible to low-income and rural populations due to a lack of physical infrastructure, internet, smartphones, and computer access. This has led to a plethora of niche players relying on innovative concepts such as Banking as a Service (BaaS) and AI capabilities to overcome these barriers and reduce disparities in financial service access between the rich and the poor.

Humanizing Financial Services with AI

AI is not only democratizing financial services but also humanizing them. AI is making financial services more customer-centric by providing personalized and targeted offers for customers. AI chatbots provide customers with more efficient and personalized customer service, making interactions with financial institutions more human-like. Automation enabled by AI can streamline processes and improve financial service efficiency, further enhancing the customer experience.

However, the humanization of financial services through AI has its challenges. AI models must be explainable to ensure fairness, privacy, and security. This is especially important when AI makes decisions that affect individuals, such as credit scoring. AI models must also be proven trustworthy for use in the financial system. The better everyone understands AI models, the more we can trust in appropriate deployment, privacy protection, and avoiding discrimination.

In an interview with the founder of Margaris Ventures, the founder, Spiros Margaris, described the humanizing efforts and effects succinctly. "Much work remains to keep educating people and customers about the vast benefits of such complex technology. We must ensure that people believe and understand that AI will benefit them as it reaches its fullest potential. Trust is still the core DNA of any business model, including banks. Therefore, as AI continues to humanize financial services, it is crucial to maintain and build on this trust (McKendrick & Margaris, 2023)."

Regulatory Compliance and AI

Regulatory compliance is a significant challenge for financial institutions. The financial services industry is highly regulated, and institutions must comply with various rules and regulations. Non-compliance can result in hefty fines and damage to an institution's reputation. AI can help financial institutions navigate the complex regulatory landscape more effectively. AI can help institutions understand and comply with regulatory requirements by analyzing regulatory texts and other relevant data.

AI can also help financial institutions manage their compliance risk more effectively. By analyzing vast amounts of data, AI can identify patterns and trends to help institutions identify potential compliance risks. For example, AI can help institutions identify transactions that may violate anti-money laundering regulations.

However, the use of AI in regulatory compliance has its challenges. AI models must be trained on vast amounts of data, which can be costly and time-consuming. This opens the door for third-party players to provide these services. Additionally, AI models must be explainable to ensure fairness and transparency. This is especially important when AI makes decisions that affect individuals, such as determining whether a transaction is compliant or vetting a customer through a know-your-customer (KYC) process.

Another challenge for deploying AI models is data storage and access, especially in light of European regulations like the General Data Protection Regulation (GDPR). Adequate security measures are necessary to ensure the safety and integrity of AI-based models.

Generative AI in Financial Services

Generative AI is making significant strides in the financial services sector. It is not just a new technology; it is a new way of thinking about and interacting with technology. Generative AI is flexible, adaptable, and capable of tasks that traditional AI systems struggle with. It is not about replacing existing AI systems but rather complementing them, filling in the gaps where traditional AI falls short. This is particularly relevant for financial services, where precision and accuracy are paramount.

Generative AI is not just about creating new models or algorithms; it is about integrating these capabilities into existing workflows and processes. For instance, generative AI can enhance customer service by providing more accurate, responsive, and resonant answers. It can also help financial analysts by simplifying complex financial data and trends. In essence, generative AI is a tool that can help financial services organizations deliver better, higher quality, and less expensive experiences for their customers.

The technology, however, is still new, and there are questions about its precision and accuracy, especially in a highly regulated industry like finance. However, as the technology matures and more use cases emerge, it is clear that generative AI has the potential to revolutionize the financial services sector.

BloomberGPT

Bloomberg has developed an AI model called BloombergGPT that uses techniques similar to those used by OpenAI's ChatGPT. This model was trained on various finance-specific sources, including financial news, company financial filings, press releases, and Bloomberg News content, as well as general datasets like The Pile, The Colossal Clean Crawled Corpus (C4), and Wikipedia (Haas & Gilmore, 2023).

The training of the BloombergGPT model required approximately 53 days of computations run on 64 servers, each containing 8 NVIDIA 40GB A100 GPUs, in partnership with NVIDIA and Amazon Web Services. The data was tokenized into 363 billion tokens using a Unigram model.

Potential applications for BloombergGPT include generating initial drafts of SEC filings, summarizing financial content into headlines, providing company organizational charts and linkages, automating the generation of draft routine market reports, and retrieving specific elements of financial statements for specific periods via a single prompt. Currently, the BloombergGPT model is not publicly available, and there is no API or chat interface to access it.

Conclusion

"With financial institutions and FinTech companies already deploying AI to improve services and stay competitive, the more tech-savvy players in the industry are likely better positioned to take full advantage of the vast possibilities and become more successful (McKendrick & Margaris, 2023)." The key financial data providers, such as Bloomberg, S&P Global Market Intelligence, and Refinitiv (formerly Thomson Reuters Financial & Risk), are well-positioned to leverage AI to solidify their market share. From employing AI capabilities in data sourcing, cleansing, and manipulation to generating insights, critical knowledge, and what-if scenario insights, AI will play a critical role in transforming financial services offerings for individuals and institutions ("Financial Data Service Providers," 2023).

Chapter 14

AI IN EDUCATION AND LEARNING

Artificial intelligence (AI) is becoming increasingly prevalent in many sectors, including education. This exciting development promises to revolutionize how education is delivered and received, but it also presents challenges that must be addressed. A detailed exploration of this topic is thus necessary to understand the full extent and potential of AI's role in education.

In the past few years, AI has transformed many aspects of daily life, from shopping and entertainment to healthcare and transportation. Education is one area that stands to benefit significantly from AI technologies. By analyzing and learning from data, AI can tailor learning experiences to individual needs, potentially improving student outcomes and creating more efficient teaching practices (Louder, 2023).

However, the use of AI in education has its challenges. Data privacy, algorithmic bias, and access to resources are challenges that must be considered carefully. There is also the question of how to prepare educators and students for an increasingly AI-driven educational system (Frąckiewicz, 2023).

Despite these challenges, the potential of AI in education is vast and promises to shape the future of learning in exciting ways. This chapter aims to examine the current state of AI in education, exploring its applications, potential, and the obstacles it may face.

Current State of AI in Education

AI is a concept that has been introduced previously in education. Educators and technologists have been exploring ways to integrate AI into teaching and learning practices for years. However, it is only recently that AI technologies have advanced enough to impact the sector significantly. This section aims to shed light on the current state of AI in education and how it is being used to enhance teaching and learning experiences (Gülen, 2023).

Adaptive Learning Systems

AI introduces the concept of adaptive learning systems in education, an approach that personalizes teaching based on each learner's needs. These systems continually assess students' understanding and modify the teaching style, pace, and content accordingly. This approach makes learning more engaging, thereby enhancing students' ability to meet their learning objectives.

Adaptive learning systems have been found to improve learning outcomes significantly. They not only foster a more engaging learning experience but also provide instant feedback to both learners and educators, enabling timely interventions when necessary. Moreover, these systems promote self-paced learning, allowing students to progress at their own speed, which can reduce stress and improve comprehension.

However, implementing adaptive learning systems is challenging. Schools and educators must invest in technology and training to effectively integrate these systems into their teaching practices. Data privacy concerns are also addressed, given the amount of personal information these systems need to function effectively.

Despite these hurdles, the potential benefits of adaptive learning systems are enormous. By providing a personalized learning experience, these systems can help to improve student outcomes and make learning more enjoyable and engaging.

Intelligent Tutoring Systems

Another application of AI in education is intelligent tutoring systems. These digital tools provide personalized instruction and instant feedback to students, helping them to fill learning gaps and improve understanding. Using AI to analyze student responses, these systems can identify where students are struggling and adjust their teaching methods accordingly (Louder, 2023).

Intelligent tutoring systems represent a significant step forward in personalized learning. They can provide one-on-one instruction tailored to each student's needs, often difficult to achieve in a traditional classroom setting. These systems can be particularly beneficial for students who may need to catch up or learn at a different pace than their peers.

However, as with other AI technologies, there are challenges to consider. Intelligent tutoring systems require a significant investment in technology and training, and there are also data privacy concerns to address. Despite these issues, the potential of these systems to transform education is clear. By providing personalized, one-on-one instruction, intelligent tutoring systems can help to ensure that all students receive the support they need to succeed.

Predictive Analytics in Education

Predictive analytics is another area where AI is making a significant impact on education. By analyzing historical data, AI can predict future trends and behaviors, helping educators to make more informed decisions. This can lead to improved student outcomes and more efficient teaching practices.

In education, predictive analytics can be used in a variety of ways. For example, it can help to identify at-risk students, allowing educators to intervene before it is too late. It can also predict future performance based on past behaviors, helping educators tailor their teaching methods to each student's needs.

However, as with other AI applications, there are challenges to consider. Predictive analytics relies on large amounts of data, raising concerns about data privacy. There are also questions about the accuracy of AI predictions and the potential for algorithmic bias.

Despite these concerns, the potential benefits of predictive analytics in education are clear. By providing insights into student behaviors and performance, predictive analytics can help educators to make more informed decisions, potentially leading to improved student outcomes (Wang et al., 2023).

AI in Classroom Management

AI can provide significant advantages in managing classroom activities. From tracking attendance to overseeing classroom behavior, AI technologies can help free up educators' time, allowing them to focus more on teaching and less on administrative tasks (Frąckiewicz, 2023). Intelligent systems can monitor student engagement, alerting teachers to issues affecting learning outcomes.

AI can also offer insights into student behavior patterns, enabling the implementation of more effective classroom management strategies. For instance, AI can analyze patterns in student behavior to identify potential disruptions before they occur, allowing for proactive management of classroom dynamics. The predictive capabilities of AI can also be used to optimize seating arrangements based on students' learning styles and interpersonal relationships.

However, integrating AI into classroom management comes with its share of challenges. Data privacy is a significant concern, especially when it involves minors. Therefore, ensuring that any technology used complies with relevant privacy laws and guidelines is vital. Additionally, implementing AI technologies requires technical know-how and investment, which could pose challenges for some schools and institutions.

Real-World Case Studies

There are many real-world case studies highlight AI integration's potential benefits and pitfalls in learning environments. These instances provide invaluable insights into the practical aspects of AI implementation, offering lessons that can be applied to future endeavors.

One example is Georgia State University's use of predictive analytics. The institution implemented an AI system to analyze student data and identify at-risk students. This approach led to a significant reduction in student dropouts and a noticeable improvement in graduation rates (Dimeo, 2017).

In another case, Carnegie Learning developed an intelligent tutoring system for mathematics. The system adapts to individual student's needs, offering personalized learning pathways and timely feedback. Schools using this system reported improved student math performance, underscoring AI's potential for personalized learning ("One-on-one Math Coaching," 2023).

However, these successful implementations come with challenges. Issues surrounding data privacy, technical requirements, and the need for ongoing training and support have often been reported. Therefore, these real-world cases serve as reminders of the need for a balanced approach, considering both the potential benefits and challenges of AI in education.

Examples of Successful Implementations

AI applications are transforming education in many ways, and several successful implementations underscore their efficacy. Pearson's AI essay scoring was one of the early examples, launching in 1998 (Pearson, 2023). The technology, Intelligent Essay Assessor, uses machine learning to assess and provide feedback on student essays, enabling timely feedback even in large classes.

Another example, Duolingo's AI-powered app has gained widespread acclaim for its innovative approach to language learning (Marr, 2020). Duolingo uses AI algorithms to personalize lessons based on learners' proficiency level and learning speed. Its gamified learning environment enhances engagement, making learning a new language fun and interactive (Marr, 2020).

Another impressive example is Third Space Learning, which uses AI to provide one-to-one math tutoring (Dickson, 2017). Its platform utilizes AI to analyze thousands of hours of teaching to identify what works best for individual students, delivering personalized learning experiences ("AI Reshaping Access to Learning," 2023).

Despite these success stories, it is crucial to understand that these AI applications are tools that supplement, not replace, human educators (Wang et al., 2023). Careful implementation and continuous monitoring ensure AI is used ethically and responsibly.

Challenges and Lessons Learned

AI's integration into education has its challenges. Data privacy, algorithmic bias, accessibility, and the digital divide are among the most prominent issues. Without careful handling, these issues can undermine the potential benefits of AI in education.

A key lesson is the importance of transparency and explainability in AI systems. Understanding how an AI system decides is crucial, particularly when those decisions impact a student's educational trajectory. Clear guidelines should be established to ensure that AI systems are transparent and that educators and students can interpret and understand their results.

Another essential takeaway is the need for collaboration among educators, technologists, policymakers, and other stakeholders. All parties should work together to shape AI's role in education, ensuring it is used to augment human teaching and not replace it. Only through such cooperative efforts can the benefits of AI in education be fully realized.

Personalized Learning: The AI Difference

The primary advantage of AI in education is its ability to facilitate personalized learning (Frąckiewicz, 2023). "Traditional classroom settings often follow a one-size-fits-all approach, which does not cater to individual learning needs ("AI Reshaping Access to Learning," 2023)." In contrast, AI can adapt to each learner's pace, knowledge level, and learning style, offering a tailor-made learning experience (Frąckiewicz, 2023).

AI-powered platforms can continuously analyze a student's progress, dynamically adjusting the difficulty level and type of content presented based on their performance (Harve, 2023). This real-time adaptation allows for a more nuanced understanding of each learner's strengths and weaknesses, enabling targeted interventions to enhance learning outcomes.

For instance, AI can help identify gaps in a student's knowledge that may be overlooked in a standard classroom setting. Once these gaps are identified, the AI system can provide specific resources or exercises to address them. This level of personalization is typically not achievable in traditional learning environments due to practical constraints.

The potential of AI in personalizing education is immense. However, it requires careful implementation to ensure that the technology is used ethically and responsibly. All stakeholders must work together to ensure AI-enhanced personalized learning is accessible, fair, and beneficial to all learners.

Customizing Learning Pathways

AI allows the creation of custom learning pathways, making education more flexible and personalized. Each learner has a unique learning style, pace, and ability to understand, and AI can tailor education to fit these individual needs. AI-powered systems can map a personalized learning pathway based on the learner's goals, strengths, and weaknesses ("Personalizing Education for Every Student," 2023). This approach can lead to more significant learning gains as it adapts to the learner's needs rather than forcing the learner to adapt to a rigid system.

For instance, students who struggle with mathematical concepts but excel in reading might receive more mathematical reinforcement and challenging reading materials. Such customization can empower students to learn independently and explore areas of interest more deeply.

Moreover, AI can facilitate the continuity of learning. In cases where learners need to shift educational settings - due to relocation, illness, or other disruptions - AI can maintain the learning pathway, reducing potential academic progress setbacks.

Early Intervention and Remediation

Early intervention and remediation are critical components of an effective education system. AI has the potential to enhance these aspects by detecting learning difficulties at an early stage and providing customized remediation strategies.

AI-powered systems can continuously monitor a student's performance and identify patterns indicating a learning challenge or deficit. Upon identifying a potential issue, AI can flag this to educators or even directly provide additional support to the student in real time. This allows immediate intervention, preventing minor difficulties from becoming major learning obstacles.

AI can also tailor remediation to each learner's unique needs. Instead of a generic set of remedial instructions, AI can develop a personalized strategy considering the learner's strengths, weaknesses, and preferred learning styles (Greene-Haroer, 2023). This makes remediation more effective and can significantly improve learning outcomes.

Transforming Assessment with AI

AI is transforming the nature of assessment in education. Traditional assessments often rely on one-size-fits-all tests that fail to capture a learner's full range of abilities and potential. AI has the potential to create more holistic and personalized assessments that can provide a more accurate picture of a student's capabilities.

AI-powered assessments can adapt in real time, adjusting the difficulty level and question types based on a student's responses (Louder, 2023). This makes the assessment experience more engaging for learners and yields more nuanced data about their understanding and skills.

Furthermore, AI can help move assessment from a predominantly summative process to a more formative one. By providing immediate feedback on a student's performance, AI enables learners to understand their mistakes and learn from them in real time. This continuous assessment and feedback loop can significantly enhance the learning process.

Automated Essay Scoring

The process of evaluating essays and written assignments is often time-consuming for educators. With the recent advancement of AI and learning models, automated essay-scoring systems have emerged. These AI-powered tools can evaluate and score essays based on predefined criteria, such as coherence, grammatical correctness, and use of vocabulary.

While automated essay scoring is not intended to replace human grading, it can be an efficient way to provide initial assessments, especially in large classes or online learning platforms with numerous participants. It offers students immediate feedback, enabling them to revise their work and learn from their mistakes promptly.

Notwithstanding, AI's application in automated essay scoring must be carefully managed. Although AI can check grammar and structure, it may need to grasp meaning, context, or creativity nuances fully. Thus, human judgment should remain a crucial component of the grading process.

AI in Formative Assessments

Formative assessments are important milestones in a learner's educational journey, enabling both the learner and the educator to gauge understanding and adjust teaching methods or learning strategies accordingly. AI can enhance these assessments, providing real-time data and actionable insights to optimize learning (Louder, 2023).

AI-driven formative assessment tools can track a student's progress in real-time, identifying areas of strength and weakness. These tools can also suggest personalized learning activities or resources to bridge learning gaps. Such real-time feedback helps learners understand where they stand and how to improve, fostering a more active, self-directed learning experience.

Furthermore, these tools can help teachers identify common areas of difficulty among students, informing changes in instruction or curriculum design. In this way, AI can support teachers in making more data-driven decisions and delivering more effective instruction (Greene-Haroer, 2023).

AI in Higher Education and Lifelong Learning

AI's transformative impact is not confined to K-12 education but extends to higher education and lifelong learning (Harve, 2023). With the rise of online learning and digital resources, AI can significantly enhance the learning experience for college students and adults seeking continuing education or upskilling opportunities.

In higher education, AI can be used to personalize instruction at scale, a particularly important capability given the diverse backgrounds and learning needs of college students. AI can also support research by helping scholars analyze large datasets, identify patterns, and generate insights.

AI can offer personalized learning pathways for adults engaged in lifelong learning that consider their existing knowledge, skills, and learning goals. AI can recommend relevant learning resources, track progress, and provide real-time feedback, creating a more flexible and learner-centered experience.

AI in Research

AI has found significant applications in academic and industry research. AI can manage and process vast amounts of data far beyond human capabilities, identifying patterns and connections that could be overlooked. This leads to more accurate results and insightful conclusions in research studies.

AI tools are used to conduct comprehensive literature reviews, which previously consumed considerable time and effort. Moreover, in fields such as healthcare or environmental science, AI can analyze complex datasets, thus improving the speed and accuracy of research.

However, ensuring that AI tools used in research are transparent and reliable is necessary. Researchers should understand the algorithms behind AI tools and verify the results produced by these tools. This can maintain the integrity and credibility of the research process.

Predicting Student Success and Retention

AI can play a substantial role in predicting student success and improving retention rates. AI can analyze various data, including academic performance, engagement levels, and social factors, to identify students who may be at risk of dropping out or performing poorly.

Early identification of such students allows educators and administrators to intervene with appropriate support services. This could range from tutoring or counseling services to teaching strategies or course content modifications.

In addition, AI can help institutions understand the factors that contribute to student success and refine their academic programs and student services accordingly. However, it is essential to remember that while AI can inform interventions, the human touch ultimately makes a difference.

AI in Continuing Education and Upskilling

Continuing education and upskilling have become critical in an increasingly dynamic and competitive job market. AI can significantly enhance these learning processes, providing personalized, flexible, and efficient learning experiences (Frąckiewicz, 2023).

AI can curate personalized learning pathways based on a learner's current skills, career goals, and learning style. It can also provide immediate feedback and adapt the content in real-time based on the learner's progress. This results in a more engaging and effective learning experience.

Moreover, AI can help identify emerging skills needed in the job market, guiding learners to upskill in areas in demand. However, to maximize the benefits of AI in continuing education and upskilling, learners need to be proactive and engaged, taking ownership of their learning journey.

AI in Special Education

AI can offer various benefits in special education, including personalized learning experiences, accessibility enhancements, and supportive tools. By analyzing individual student data, AI can help create customized learning pathways "tailored to each student's unique needs and abilities (Frąckiewicz, 2023)," fostering inclusion and accessibility.

AI-powered assistive technologies, such as speech-to-text and text-to-speech converters, augmentative and alternative communication (AAC) tools, and personalized learning applications, can significantly improve the learning experiences of students with special needs (Zdravkova et al., 2022).

Nevertheless, it is crucial to remember that AI's effectiveness in special education heavily depends on the context, the specific needs of the student, and the quality of the AI tools being used. Also, the ethical use of data in this context is paramount to safeguard student privacy and dignity.

Future of AI in Education

Looking ahead, AI holds the promise to revolutionize the educational sector profoundly. It could redefine how learning is delivered, assessed, and certified, making education more personalized, efficient, and impactful. Innovations that may change classrooms, curriculums, and pedagogy are being developed, including intelligent tutoring systems, personalized learning environments, and adaptive assessment tools.

However, as the applications of AI in education continue to evolve, it is critical to address challenges, including data privacy concerns, algorithmic transparency, and the risk of widening socio-economic divides in access to quality education (Frąckiewicz, 2023). Policymakers, educators, and technologists should collaborate to navigate these challenges and harness AI's potential responsibly.

Ethical Considerations

The adoption of AI in education introduces various ethical considerations. Data privacy is a primary concern, as AI systems often require extensive data about students to function effectively (Gülen, 2023). It is essential to ensure that this data is collected, stored, and used, respecting students' privacy and consent (Greene-Haroer, 2023).

Moreover, the algorithms used by AI should be transparent, interpretable, and free from bias. Decision-making processes in AI systems should be explainable, and the criteria used should not discriminate against certain student groups. Also, ensuring that the increased use of AI does not exacerbate existing inequalities in access to quality education is critical.

Preparing for an AI-Enhanced Education Future

Preparing for an AI-enhanced future in education involves several steps. First, an understanding of AI and its implications should be cultivated among educators, students, parents, and policymakers (Greene-Haroer, 2023). This can enable informed decision-making about the use of AI in education.

Second, appropriate policies and regulations should be developed to guide the ethical use of AI in education (Gülen, 2023). This includes ensuring data privacy, algorithmic transparency, and equitable access to AI-enhanced education.

Lastly, continuous research and dialogue should be encouraged to explore the potential of AI in education, identify emerging challenges, and develop effective strategies to address them (Wang et al., 2023). The goal should be leveraging AI to enhance learning while preserving the human touch at the heart of education.

Conclusion

AI in education and learning is an evolving field with the potential to redefine teaching and learning processes. As we explore and implement AI technologies in education, careful consideration should be given to maintaining the balance between technological advancement and the traditional values of education. It is a promising future, with AI amplifying the human capacity to educate and learn, fostering more personalized, accessible, and efficient educational experiences.

Chapter 15

AUTHENTICITY IN THE AGE OF AI: THE PRICE OF HUMAN TOUCH

The notion of authenticity in AI-generated content can be multilayered and complex. At its core, authenticity refers to the genuineness of the content, the sense that it has been created with intentionality, creativity, and the uniquely human ability to express nuanced emotions and thoughts. Authenticity in AI-generated content points to the degree to which the content can mirror these human attributes and how successful it is at convincing the recipient of its human-like origin ("Valuing the Human Touch", 2023). However, even as AI advances to create content that may appear strikingly similar to human output, it is worth considering if that truly encapsulates authenticity, given the lack of human experience that underpins the AI's creation (Dey, 2023).

Authenticity is not solely about mimicking human creativity convincingly. It also encompasses the ethical and moral dimensions of content creation. Authentic content is produced in an honest and transparent manner, with a clear acknowledgment of its origins. Thus, in AI-generated content, authenticity would also imply transparency about its machine origin rather than pretending to be human-made. This dichotomy presents one of the many intriguing debates in AI.

While the concept of authenticity in AI-generated content may seem abstract, it is important to wrestle with these complexities. As AI continues to permeate every facet of our lives and businesses, increasingly leveraging AI for various functions, understanding what authenticity means in this context will be crucial. It will inform how we interact with AI-generated content, perceive its value, and navigate its ethical challenges.

The Human Touch: The Core of Authenticity

Definition and Significance of the "Human Touch" in Content

Referring to the "human touch" in content alludes to the inimitable human elements that permeate our creative expressions. It is the idiosyncrasies in our writing; the emotions conveyed through a painting or the heartfelt empathy in customer service. The human touch can be found in the stories we tell, the relationships we cultivate, and the shared experiences that bring us together. This human touch often distinguishes memorable content from the forgettable, fostering an emotional connection that resonates with audiences.

The human touch goes beyond the content; it represents the creator's worldview, emotions, experiences, and values. It breathes life and authenticity into content, enabling it to transcend its functional purpose and connect with audiences on an emotional level. This personal and emotional connection often makes content engaging, compelling, and memorable.

The significance of the human touch in content cannot be overstated. In a world increasingly dominated by automated content, the human touch is the differentiating factor that can set a brand, a piece of content, or a service apart. It offers a unique value that AI cannot replicate, underscoring the importance of authenticity in an increasingly automated world.

Why the Human Touch Matters in Business and Customer Relationships

The role of the human touch in business and customer relationships is multifaceted and increasingly vital. Businesses prioritizing the human touch distinguish themselves in a marketplace teeming with AI-powered interactions, fostering deeper customer connections. Whether through personalized service, empathetic customer support, or thoughtful marketing communications, the human touch enhances the customer experience in ways that AI cannot match.

The human touch introduces an emotional dimension to business transactions, transforming them from mere exchanges of goods or services into meaningful interactions. This is particularly evident in healthcare, hospitality, and customer service sectors, where empathy, understanding, and personalization are paramount. The human touch can create positive customer experiences, foster loyalty, and build trust.

Moreover, the human touch can elevate a company's brand by imbuing it with authenticity. Customers increasingly value authentic, transparent, and socially conscious brands. By maintaining a human touch in their operations, businesses can effectively convey these qualities, enhancing their brand value and customer loyalty.

Human Touch and Technology

One might be tempted to think that the human touch is becoming obsolete. However, the reality is far from it. The human touch, characterized by empathy, understanding, creativity, and emotional intelligence, continues to be integral in building relationships, fostering customer loyalty, and driving innovation. These are aspects that AI, despite its sophisticated algorithms and learning capabilities, is yet to replicate fully (Lee, 2023).

The human touch brings an emotional depth to interactions, an ability to understand and respond to non-verbal cues, empathize with the customer's situation, and make judgments that may defy rigid, pre-programmed rules but make perfect sense in human relations. In a business environment, this translates to better customer service, more effective negotiation, and more innovative problem-solving. As much as AI can support and augment these functions, the human touch remains a differentiator, adding a layer of value that is uniquely human ("Valuing the Human Touch", 2023).

Moreover, in technology, the human touch is essential in designing and implementing solutions. Technology, after all, is built to serve human needs. Understanding those needs, empathizing with the end-user, and being able to envision solutions from a human perspective are key aspects that drive successful technological innovations. This is where the human touch comes into play, guiding the application of technology in a meaningful, beneficial, and ethical way.

The Role of Human Emotions in Making Business Decisions

Human emotions play a pivotal role in business decision-making. Unlike AI, humans are not purely rational actors; many emotional factors, including our values, aspirations, fears, and desires, influence our decisions. This also holds true in business, where decisions often involve weighing various emotional and rational factors.

In marketing and sales, understanding the emotional underpinnings of consumer behavior can unlock powerful strategies. Emotionally resonant advertising, empathetic customer service, and products that cater to consumers' emotional needs can significantly influence purchasing decisions, driving revenue and market share.

Moreover, emotions can drive innovation and strategic decision-making within organizations. Passion for a product, mission, or customer base can inspire innovative ideas and bold strategic moves. Likewise, empathy for customers or employees can lead to business decisions that enhance satisfaction and loyalty, driving long-term success.

The Economic Value of Authenticity in an AI-Driven World

Understanding the Concept of Authenticity Premium

In an AI-driven world, there is a potential for a new kind of value to emerge: the "authenticity premium (Buhr et al., 2021)." This refers to the added value or worth that consumers assign to goods, services, or content perceived as genuine, original, or human-produced instead of machine-generated (Buhr et al., 2021). As AI-generated content becomes increasingly pervasive, consumers may place a higher value on human-created content, viewing it as a luxury or rarity.

The authenticity premium stems from the unique qualities that human creativity brings – the emotional depth, the personal touch, the unpredictability, and the capacity for empathy (Buhr et al., 2021). These are attributes that, despite major strides in AI, still remain distinctively human. As a result, consumers may be willing to pay a premium for products and services that offer these human qualities.

However, the authenticity premium is not just about humans vs. machines. It also involves the degree of transparency and honesty with which businesses operate. In an age where AI can generate highly realistic content, businesses that are upfront about using AI and use AI ethically and responsibly may also command an authenticity premium.

The Potential Impact on Different Sectors

The authenticity premium could have a significant impact on a variety of sectors (Buhr et al., 2021). In the creative industries, for instance, art, music, literature, and other forms of human expression could be more highly valued if they are perceived as authentic human creations. This does not necessarily mean that AI-generated content will not have a market. Indeed, it can cater to a separate demand for affordable, accessible, and customizable content. However, alongside it, there may be a parallel market for authentic human-created content, which commands a higher price due to its perceived rarity and uniqueness (Buhr et al., 2021).

In the service industry, a similar dynamic could play out. AI can undeniably enhance the efficiency, personalization, and accessibility of services. However, services that involve a high degree of human interaction, emotional understanding, and personal touch could become more valuable. This could range from sectors like healthcare and counseling, where empathy and human connection are critical, to sectors like hospitality and retail, where personal service can significantly enhance customer experience.

Implications for Businesses and Technology Leaders

For businesses and technology leaders, the emergence of an authenticity premium presents both challenges and opportunities. On the one hand, it reinforces the importance of human creativity and interaction, indicating that AI cannot completely replace these. This could mean investing in training and development to enhance these uniquely human skills within the workforce or designing services and products that highlight human involvement (Whitfield, 2021).

On the other hand, the authenticity premium also signals the need for ethical and transparent use of AI. Businesses that use AI must be upfront about it and must use it in a way that respects user privacy and autonomy. This could involve implementing robust ethics policies, investing in AI transparency, or even engaging in third-party audits of AI usage. This helps earn consumer trust and positions the business as a responsible player in the AI domain (Whitfield, 2021).

The authenticity premium also presents an opportunity for differentiation and value creation. Businesses can leverage it to create unique offerings that combine the best of human creativity and AI efficiency. This could involve, for instance, personalized services that use AI for backend operations but involve human interaction in the customer-facing aspects. Alternatively, it could involve products designed by humans but customized by AI, offering a balance of authenticity and personalization. Ultimately, navigating the authenticity premium will require a nuanced understanding of the interplay between AI and human creativity and a strategic approach to integrating the two.

The Potential Economic Value of Authenticity in an AI-Dominated World

As AI continues to permeate our lives and businesses, the economic value of authenticity is poised to rise. In an AI-dominated world, authenticity — the human touch — can become a scarce and, therefore, valuable commodity. Businesses that successfully incorporate this human element into their offerings can differentiate themselves in the market and command a premium for their services.

The economic value of authenticity can manifest in several ways. For one, it can drive customer loyalty and retention. Customers will likely stay loyal to brands they perceive as authentic and consistent in their values. "This loyalty can translate into repeat business, positive word-of-mouth, and a stronger brand reputation ('Power of Storytelling,' 2023)."

In addition, authenticity can also attract new customers. As consumers become more discerning and demand more than just functional value from their purchases, businesses that demonstrate authenticity can appeal to these consumers' desire for meaningful, authentic experiences.

Finally, authenticity can drive innovation. By staying true to their mission and values, businesses can foster a culture of authenticity that encourages innovation. This can result in unique, differentiated products and services that command a premium in the market.

The Willingness of Consumers to Pay for "Human-Authentic" Services and Content

Research and market trends indicate a growing willingness among consumers to pay a premium for "human-authentic" services and content. Amid the increasing deluge of AI-generated content and automated services, consumers will value the uniqueness, creativity, and personal touch that only humans can provide (Whitfield, 2021).

A clear indicator of this trend is the resurgence of artisanal and handcrafted products in various industries, from food and beverages to fashion and home décor (Helmore, 2022). Despite the availability of mass-produced alternatives, many consumers are willing to pay a premium for products that embody their creators' craftsmanship, creativity, and personal touch.

The same applies to services, particularly in the healthcare, education, and customer service sectors. In these sectors, the human touch — characterized by empathy, understanding, and personal attention — is highly valued by consumers, often commanding a higher price than automated alternatives.

Furthermore, there is a growing appetite for unique, human-centric content in content creation. Whether it is storytelling in marketing, thought leadership in business, or entertainment in media, content that showcases human creativity, perspectives, and emotions is highly sought after, further underscoring the economic value of the human touch.

Case Studies of Businesses Profiting from Human Touch in an Increasingly Automated World

Several businesses have successfully leveraged the human touch to differentiate themselves in an increasingly automated world, reaping substantial profits. These case studies underscore the economic potential of authenticity and the human touch.

One such example is the rise of direct-to-consumer (D2C) brands, many of which have built their success on a foundation of authenticity and human connection. Brands like Glossier in cosmetics or Allbirds in footwear have differentiated themselves through authentic storytelling, community engagement, and a commitment to sustainability and inclusivity (Tran, 2022). This human touch has resonated with consumers, driving customer loyalty and enabling these brands to command a premium for their products. It is worth noting that these two brands have capitalized on their success and moved into physical retail stores ("Omnichannel Evolution," 2022).

In the hospitality sector, businesses like Airbnb have capitalized on the human touch to disrupt traditional business models. By facilitating unique, local experiences and personal connections between hosts and guests, Airbnb has created a differentiated offering that has resonated with travelers seeking authentic experiences (Gallagher, 2018).

Finally, in the customer service sector, companies like Zappos have become renowned for their exceptional, human-centric customer service. Zappos has fostered a loyal customer base and a strong brand reputation by empowering its service representatives to go above and beyond to satisfy customers (Ordoñez, 2018).

These examples illustrate how the human touch can drive economic value, even in an AI-dominated world. They highlight the importance of authenticity in business strategy and underscore the potential rewards for businesses that can successfully incorporate the human touch into their operations.

AI and the Paradox of Authenticity

The Rise of AI-Generated Content: An Overview

AI has firmly entrenched itself as a paradigm-shifting force in the world of content creation. Leveraging its unprecedented computational capabilities and the insatiable appetite for data, AI's rise as a powerful content generator has been meteoric. AI is now capable of creating diverse content, from compelling narratives and insightful articles to mesmerizing artwork and captivating music, across a multitude of domains.

The term "AI-generated content" encapsulates creations produced with minimal to no human intervention, with AI systems harnessing vast data lakes to understand, learn, and eventually mimic the intricacies of human expression. These AI systems are programmed with machine learning algorithms capable of identifying patterns, trends, and relationships in the data they analyze, extrapolating from them to generate real-time content. The ever-expanding capabilities of AI-generated content are a double-edged sword. They offer incredible possibilities and efficiencies while raising profound questions about authenticity, creativity, and the intrinsic value of human endeavor.

How AI-Generated Content Often Lacks "Authenticity"

Despite the prowess of AI in generating content, a key facet often needs to be discovered - authenticity. By definition, AI systems are devoid of lived experiences, emotions, and subjective interpretations of the world, which form the bedrock of human creativity. While these systems can master the mechanics of language or design principles, they cannot understand the humanistic subtleties imbued in these creative expressions.

While increasingly sophisticated, AI-generated content often appears formulaic and lacks the "soul" that characterizes human creativity. The emotionally resonant nuances, the individual idiosyncrasies, and the unpredictability that breathe life into human creations are inherently absent in AI-produced content. This absence is most noticeable in creations requiring a deep understanding of human emotions and experiences, such as literature, poetry, or personalized customer interactions.

Moreover, AI operates within the boundaries of its programming and the data it has been fed. It mirrors what it has learned rather than creating something entirely novel or groundbreaking. While AI can undoubtedly generate content at a scale and speed beyond human capability, the value of such content may be undermined if it is perceived as inauthentic or soulless. The lack of authenticity in AI-generated content is a paradox we must grapple with as we continue to leverage this technology.

The Paradox: AI's Goal to Emulate Human Intelligence vs. the Inherent Uniqueness of Human Touch

The rise of AI and its increasing role in content creation has engendered a paradox that presents a fascinating conundrum. On the one hand, AI's fundamental aim, particularly in content generation, is to emulate human intelligence to such a degree that its creations are indistinguishable from those produced by humans. On the other hand, this very emulation reveals the inherent uniqueness of the human touch, underlining the intangible qualities that set humans apart.

In its quest to replicate human intelligence, AI is trying to capture a moving target. Human intelligence is not static; it evolves, adapts, and matures. It is influenced by our experiences, emotions, cultural contexts, and many other factors that contribute to our unique perspectives and creative expressions. It is an amalgamation of conscious and unconscious thoughts underpinned by deep-seated emotional undercurrents and nuanced subtext, all intrinsically human.

Furthermore, human creativity is not merely the act of producing something new. It is deeply intertwined with our sense of self, values, and emotional responses. These aspects are incredibly challenging, if not impossible, for AI to replicate. While AI can convincingly simulate human-like content, it cannot recreate the human experience that underpins our creative endeavors. This paradox, wherein AI's success in emulating human intelligence underscores the inherent uniqueness of human creativity, is one of the most intriguing aspects of our ongoing journey with AI.

In essence, the role of human emotions in business is profound. Understanding and leveraging these emotional dynamics could be key to maintaining a competitive edge as we navigate AI-dominated services.

The Future of Authenticity in AI

Predictions for the Evolution of AI and its Ability to Mimic Human Authenticity

The pace of AI advancement is staggering. AI's anticipated to continue to evolve, improving its ability to mimic human-like attributes, including creativity, empathy, and even the human touch. However, despite these advancements, AI will only partially replicate human authenticity's richness, depth, and complexity.

The reason for this lies in the inherent nature of authenticity. Authenticity is intrinsically linked to our lived experiences, emotions, values, and idiosyncrasies. It is shaped by our unique life histories, cultural backgrounds, and personal beliefs. While AI can simulate certain aspects of human behavior and even mimic emotions, it cannot replicate the full range of human experiences and the personal growth that comes from living those experiences.

Moreover, the authenticity of the human touch also stems from its unpredictability and spontaneity. It is in the unexpected insights during a brainstorming session, the impromptu acts of kindness, or the unique stories that only humans can tell. Given its inherent dependence on predefined algorithms and data, these aspects are challenging for AI to replicate.

However, AI will still play a significant role in content creation and services in the future. As AI becomes more sophisticated, we can expect more AI-generated content that closely mimics the human touch, blurring the lines between human and AI-generated content. However, even then, the truly authentic human touch will likely retain a unique value.

Ethical and Societal Implications of Indistinguishable AI-Generated Content

As AI advances and its ability to mimic the human touch improves, several ethical and societal implications emerge. One major concern is the potential for AI-generated content to be used deceptively, blurring the lines between human and AI-generated content and misleading consumers. This raises questions about transparency, consent, and the right to know the true origin of content.

Moreover, the rise of sophisticated AI-generated content could have profound implications for jobs and livelihoods. If AI can create content that closely mimics the human touch, it could displace humans in roles traditionally requiring creativity and human interaction. This could lead to job losses and widen the economic divide between those who can adapt to the AI-dominated world and those who cannot.

On a societal level, the proliferation of AI-generated content could also contribute to a sense of alienation and dehumanization. If AI increasingly mediates our interactions, entertainment, and information, we may find ourselves longing for the authenticity and connection that only human-generated content can provide.

Despite these potential challenges, the rise of AI could lead to a newfound appreciation for the human touch. Just as the industrial revolution led to a resurgence in handcrafted goods, the AI revolution could spark a renewed desire for human-generated content and services, underscoring the enduring value of the human touch.

Strategies for Businesses to Maintain Authenticity in an AI-Driven World

When navigating this new AI-dominated world, businesses must consider how they can maintain authenticity while leveraging the benefits of AI. Several strategies can help businesses strike this balance.

First, businesses can prioritize transparency in their use of AI. By clearly communicating when and how AI is used, businesses can build trust with consumers and ensure they are not misled by AI-generated content (Brodsky, 2022).

Second, businesses can ensure that AI enhances the human touch rather than replacing it. For example, AI can automate repetitive tasks, freeing humans to focus on areas where they can add unique value, such as strategic decision-making, creative tasks, or customer interactions.

Third, businesses can foster a culture of authenticity, emphasizing values like empathy, integrity, and social responsibility (Bolden et al., 2023). By fostering this culture, businesses can ensure that their human touch remains at the forefront, even as they adopt AI technologies.

Finally, businesses can invest in reskilling and upskilling their workforce, preparing them for an AI-dominated world (Bolden et al., 2023). By focusing on skills that AI cannot replicate, such as emotional intelligence, creativity, and strategic thinking, businesses can ensure they continue to offer a unique human touch.

Preparing for an AI-Dominated Future: Recommendations for Business Leaders

Balancing AI Adoption with the Need for Authenticity

Balancing AI adoption with the need for authenticity is a critical task facing business leaders today. As leaders navigate this delicate balance, they must keep in mind that AI should be a tool that enhances human capabilities, not a replacement for the human touch.

To strike this balance, leaders should first consider where AI can add the most value. In many cases, AI can automate repetitive tasks, analyze large datasets, and improve efficiency. By using AI in these areas, businesses can free up human resources to focus on tasks that require a human touch, such as building customer relationships, creating strategic plans, or developing creative content (Bolden et al., 2023).

Leaders should also consider the ethical implications of AI use. As AI becomes more sophisticated, issues of transparency, privacy, and fairness become increasingly important. Leaders must ensure that their use of AI aligns with their company's values and societal expectations, reinforcing their authenticity.

Finally, leaders should regularly reassess their AI strategy. As AI continues to evolve, new opportunities and challenges will arise. By staying adaptable and keeping the human touch at the forefront of their strategy, leaders can ensure they reap the benefits of AI without sacrificing authenticity.

Nurturing Human Skills that AI Cannot Replicate

In an AI-dominated world, human skills that AI cannot replicate are becoming increasingly valuable. These skills often revolve around creativity, emotional intelligence, and complex problem-solving.

Creativity is one area where humans have a distinct advantage. While AI can generate content based on existing data and patterns, it needs help producing novel ideas. In contrast, humans can draw on a wealth of experiences and emotions to create unique, innovative content (Cerejo, 2023).

Emotional intelligence is another unique human skill. While AI can mimic certain emotional responses, it cannot fully understand or replicate human emotions. Emotional intelligence, which involves understanding and responding to the emotions of others, is critical in roles involving human interaction, such as leadership, customer service, and sales.

Complex problem-solving is a third area where humans excel. Humans can consider many factors, including emotional and ethical considerations when solving problems. This ability to navigate complexity and ambiguity is invaluable in strategic decision-making and leadership roles.

To nurture these skills, businesses can invest in training and development programs, create a culture that values and rewards these skills, and design roles that allow these skills to shine.

Establishing Guidelines for AI Usage to Preserve Authenticity

As businesses increasingly adopt AI, establishing guidelines for its usage can help preserve authenticity. These guidelines should outline how AI should be used, the ethical considerations to consider, and the steps to ensure transparency and accountability (Lee, 2023).

For example, guidelines could stipulate that AI should be used to enhance the human touch, not replace it. This could involve using AI to automate repetitive tasks or analyze data, freeing humans to focus on tasks requiring emotional intelligence, creativity, and complex problem-solving.

Ethically, guidelines should address issues like fairness, privacy, and transparency. They should ensure that AI is used in a way that respects individual rights, avoids bias, and is transparent to consumers and stakeholders (Lee, 2023).

Finally, guidelines should outline processes for auditing and reviewing AI systems to ensure accountability. This could involve regular audits to check for bias or errors and reviews to assess the impact of AI on employees, customers, and the business as a whole.

Conclusion

As we move towards an increasingly AI-dominated future, the human touch and authenticity have emerged as potent differentiators in the business world.

While AI brings significant benefits, from efficiency gains to powerful insights, it cannot replicate the human authenticity's depth, complexity, and richness, authenticity, imbued with our unique experiences, emotions, and values, is a powerful tool for businesses, fostering deeper connections with customers, driving innovation, and enhancing brand value.

As business leaders, striking the right balance between AI adoption and maintaining authenticity is crucial. This involves using AI to enhance, not replace, the human touch, nurturing human skills that AI cannot replicate, and establishing guidelines for ethical and transparent AI use.

In the AI-dominated world of tomorrow, authenticity will not only be a key differentiator and an economic driver. Businesses that can harness the power of the human touch while effectively leveraging AI, are poised to thrive in this brave new world.

Chapter 16

FUTURE OF AI: THE PATH AHEAD

AI continues to change the shape of business as we know it. AI models, like ChatGPT from OpenAI, are capable of tasks traditionally tied to "knowledge work," such as customer service, news reporting, legal document creation, and copywriting. These capabilities have sparked discussions on job displacement. However, as AI evolves, it is not just about human tasks being replaced by machines. Instead, we are witnessing the creation of entirely new systems, jobs, and business models enabled by AI technologies. For instance, ride-hailing services like Uber and Lyft leveraged AI to revolutionize taxi dispatch systems, spawning an entirely new ride-service model.

Likewise, AI's ability to generate human-like text could democratize quality writing, spurring the development of new systems and business models. AI's potential to generate code could dramatically reshape the software development industry. Simultaneously, AI-powered graphic design tools like DALL-E democratize image creation, potentially transforming industries from education to marketing.

While these advances promise to unlock countless opportunities, they also risk disrupting the economic environment. The associated "adjustment costs" of job displacement cannot be ignored, and it is essential to think about how these new capabilities can be harnessed for societal benefit.

Leveraging AI for Enhanced Strategic Planning

Strategic planning is vital for businesses to thrive in the dynamic and unpredictable world. Integrating AI into strategic planning processes can significantly enhance an organization's ability to make informed decisions, adapt to changing circumstances, and identify new growth opportunities.

Real-Time Insights: AI-powered analytics can provide real-time insights by analyzing vast amounts of data from internal and external sources. This enables organizations to comprehensively understand market trends, customer behavior, and competitive dynamics, empowering them to make data-driven decisions promptly.

Continuous Monitoring and Adaptation: AI can continuously monitor the usage of knowledge management systems, productivity tools, and other internal and external systems, collecting data and detecting patterns, enabling organizations to adapt their strategic plans in response to changing conditions. Organizations can remain agile and responsive to emerging opportunities or risks by dynamically updating plans based on new information.

Identifying Disruptions and Opportunities: AI can help businesses identify potential disruptions and emerging trends by analyzing data from various sources, such as social media, industry reports, and customer feedback. This allows organizations to proactively capitalize on emerging opportunities or mitigate potential threats, giving them a competitive edge. What was once done through a manual SWOT analysis can now be deduced and presented by AI with high accuracy.

Scenario Planning and Analysis: AI can facilitate scenario planning by simulating future scenarios based on assumptions and data inputs. Through AI-powered modeling and simulation, organizations can assess the potential impact of different strategies, evaluate risks, and optimize decision-making under different scenarios. This enables more robust strategic planning and minimizes the effects of uncertainties.

By incorporating AI into strategic planning, organizations can enhance their ability to make data-driven decisions, adapt to changing market conditions, and seize new opportunities. However, it is important to remember that AI is a tool that should be used with human expertise and judgment. Strategic planners should carefully interpret AI-generated insights and ensure alignment with their organization's values, goals, and objectives. With AI as a strategic planning tool, organizations can navigate complexities, uncover hidden insights, and stay ahead in a rapidly evolving business environment.

Strategy Development

Armed with these insights, AI can enhance the quality of decisions and shape better business strategies through its various stages of strategic development.

Descriptive Intelligence - Interpreting What Happened: AI helps consolidate and visualize complex data at the simplest level. This descriptive intelligence capability allows AI to create dynamic dashboards, track real-time performance metrics, and provide comprehensive views of business and market conditions. This aids in competitive analysis, historical performance tracking, and identifying trends or patterns. This is extremely helpful in operational areas of focus, especially IT operations, where millions of telemetry data must be visualized and represented.

Diagnostic Intelligence - Understanding Why It Happened: Moving one step further, AI helps diagnose why certain events occurred. By applying machine learning algorithms to historical data, diagnostic intelligence can reveal past performance's root causes and drivers. This analysis can help businesses understand the reasons behind their successes or failures and accordingly adjust their strategies.

Predictive Intelligence - Forecasting What Might Happen: AI's predictive intelligence capabilities allow businesses to anticipate future scenarios based on past trends and current market signals. This can involve forecasting sales, predicting customer behavior, or identifying potential risks. These insights enable businesses to plan proactively and make forward-thinking decisions.

Prescriptive Intelligence - Advising on What Should Be Done: At a more advanced level, AI can suggest data-driven actions to achieve specific objectives. Using complex algorithms, prescriptive intelligence can analyze multiple variables and outcomes to recommend the best action. This can involve advising on optimal pricing strategies, resource allocation, or marketing campaigns, helping businesses optimize performance and create value.

Delegated Decision-Making - Making Decisions with Human Supervision: As AI technologies advance, they can be assigned limited decision-making authority, albeit with human oversight and defined constraints. For instance, AI systems can be programmed to automate certain operational decisions, such as adjusting digital ad spend, scaling computing capabilities in a cloud environment or recommending personalized product offerings to customers.

Fully Autonomous AI - Making Strategic Decisions Independently: The eventual, albeit ambitious, goal is to develop AI systems capable of independently analyzing and making strategic decisions. While this level of autonomy is still in its nascent stages and carries significant ethical and governance implications, it represents the future potential of AI in strategy development. It is worth noting that an autonomous generative pre-trained model, GPT, has been configured to pursue novel outcomes, such as negotiating a phone bill reduction, managing a social media channel, or placing your Instacart orders (Garg, 2023).

By integrating AI across these stages, businesses can enhance their strategic decision-making capabilities, driving improved performance and competitive advantage. However, it is crucial to maintain a balanced approach, combining the power of AI with human expertise and judgment, to ensure responsible and effective strategic decisions.

Workplace and AI

Rather than replacing jobs outright, AI will likely change the nature of many professions. AI tools can help approach business problems innovatively, reshaping careers and workplaces in ways we cannot currently foresee. Automating and creating new tasks will be integral to this transformation, necessitating professionals to adapt their roles and responsibilities.

Integrating AI into jobs might increase productivity and redefine work for many. In this emerging AI era, key skills include strategic thinking, creativity, adaptability, emotional intelligence, and leadership. "First-draft automation," where AI generates initial drafts of documents ranging from press releases to legal contracts, is an area of notable impact, extending to 3D modeling and code drafting. This shift will likely boost demand for professionals adept at using AI for text, code, and model generation and those capable of integrating these results into broader systems. However, professionals need not be AI experts. Understanding AI's fundamentals, potential, and limitations is enough to collaborate with technical teams effectively, implement AI solutions, and incorporate AI into business strategies.

Emerging Trends in AI

AI's evolution continues to produce innovative trends and technologies, molding the trajectory of business strategy. By 2030, the AI market is expected to exceed \$2 trillion, reflecting the increasing integration of AI across various industries ("Fortune Insights", 2023). This emphasizes the importance for businesses to stay updated with these advancements, leverage AI's potential, and sustain competitiveness fully. Here are some significant trends to watch out for:

Generative AI

Generative AI, AI's capability to generate novel content, has gained impressive attention. This technology involves algorithms creating new content from existing data, such as text, images, or sounds. The success of OpenAI's GPT-3, capable of generating human-like text, typifies this trend. With the power to mimic human creativity, Generative AI has extensive implications, from producing synthetic data for businesses to creating deep fake videos for entertainment.

Ethical and Explainable AI

Ethics and explainability take center stage as we increasingly rely on AI for decision-making. AI systems must operate transparently and fairly, especially when handling sensitive personal data. Efforts to solve the "black box" problem, where the decision-making process of AI systems is obscure, are gaining momentum in 2023 (Alteryx, 2023). Explainable AI, also known as XAI, is gaining prominence to make AI models and algorithms more transparent and interpretable, enabling organizations to understand the rationale behind AI-generated recommendations and decisions. Explainable AI improves trust and accountability and helps organizations address regulatory requirements and ethical considerations.

Augmented Working

The integration of AI into the workplace is revolutionizing how we work. Intelligent machines and AI-powered systems increasingly collaborate with humans, enhancing our capabilities and efficiency. Whether through smart handsets providing immediate access to data, AR-enabled headsets overlaying digital information onto the physical world, or AI assistants automating routine tasks, AI is becoming an indispensable component of various professions. This topic will be discussed in more detail later in the chapter.

Sustainable AI

With growing environmental concerns, the sustainability of AI practices is under scrutiny. AI algorithms and the supporting infrastructure require significant energy resources, resulting in substantial carbon emissions. However, AI also presents opportunities to help companies understand and reduce waste and inefficiency, driving more sustainable practices. Efforts are in progress to balance the environmental costs and benefits of AI, emphasizing utilizing green and renewable energy sources (Wynsberghe, 2021).

Edge Computing and AI

Edge computing, which involves processing data closer to the source rather than relying on centralized cloud servers, is gaining traction in conjunction with AI. Edge computing enables real-time data analysis, reducing latency and enabling faster decision-making. This is particularly important in scenarios where real-time responses are crucial, such as autonomous vehicles, industrial IoT, and remote healthcare monitoring. Combining edge computing and AI empowers organizations to leverage AI capabilities directly on edge devices, enabling faster and more efficient AI-powered applications (Hua et al., 2023).

Federated Learning

Federated learning is an emerging approach that allows multiple devices or systems to train a shared AI model collaboratively while keeping the data decentralized. This technique enables organizations to leverage collective knowledge from various sources without compromising data privacy and security. Federated learning is particularly valuable when data privacy regulations restrict centralized data aggregation. It allows organizations to build robust AI models while maintaining data privacy and ownership.

AI-Enabled Robotics

Integrating AI with robotics is revolutionizing manufacturing, logistics, and healthcare industries. AI-enabled robots can perform complex tasks with increased precision, efficiency, and adaptability. These robots can leverage AI capabilities, such as computer vision and natural language processing, to interact with the environment and collaborate with humans effectively. The combination of AI and robotics presents significant opportunities for automating labor-intensive processes, enhancing operational efficiency, and enabling human-robot collaboration.

Quantum Computing and AI

Quantum computing has the potential to revolutionize AI by exponentially increasing computational power and solving complex problems that are currently infeasible for classical computers. Quantum machine learning algorithms can leverage the unique properties of quantum systems to perform more efficient optimization, pattern recognition, and data analysis. Although quantum computing is still nascent, it promises to unlock new possibilities for AI, particularly in drug discovery, materials science, and optimization problems.

Human-Centered AI

The focus on human-centered AI is growing, emphasizing the development of AI systems that are ethical, unbiased, and aligned with human values. Human-centered AI considers AI's social and ethical implications and places humans at the center of AI development and deployment. It involves ensuring fairness, transparency, and accountability in AI algorithms, addressing biases, and promoting inclusivity. Human-centered AI also encompasses the responsible use of AI, ensuring that AI technologies benefit society.

AI in the Public Domain

The rise of AI and its applications have had significant impacts on democracy, particularly in three main areas: the public sphere, elections, and public services:

AI and Democracy: AI is changing the dynamics of the public sphere, particularly with the rise of social networks. The personalization of news consumption and the blurring of lines between private and public conversations can lead to an echo chamber or filter-bubble effects, which are thought to intensify political polarization. The application of AI is seen to enhance the possibilities for analyzing and steering public discourses, and it can boost the automated compartmentalizing of will formation. The risk emerging from this development is twofold. On the one hand, malicious actors can use these new possibilities to manipulate citizens on a massive scale, as seen in the Cambridge Analytica scandal. On the other hand, the changing relationship between public and private corporations raises issues about opaque influences over political processes and accountability (Thiel, 2022).

AI and Elections: AI technologies can manipulate public opinion and voter behavior. A notable example was the Cambridge Analytica scandal, where data was harvested from millions of Facebook profiles and used to influence the 2016 U.S. Presidential Election. AI can be used for microtargeting, which involves sending tailored messages to specific individuals or groups to influence their behavior. Concerns exist that such practices undermine democratic principles as they could manipulate people's opinions and behaviors without their knowledge or consent. However, there is an ongoing debate about the actual effectiveness of such practices and their impact on the democratic process (Thiel, 2022).

AI and Public Services: AI technologies are increasingly used in the public sector to automate decision-making processes, from social services to law enforcement. While AI can increase efficiency and effectiveness, there are concerns about accountability, fairness, and transparency. Automated decision-making systems can perpetuate biases present in the data they were trained on, which could lead to discriminatory outcomes. Moreover, the lack of transparency and understanding of how these systems make decisions can undermine public trust and accountability (Thiel, 2022).

To address these challenges, it is essential to have comprehensive regulatory frameworks that consider the political dimensions of AI and uphold democratic principles. The regulation should ensure the fair and transparent use of AI, prevent malicious uses, and ensure that private entities are held accountable for using AI (Thiel, 2022).

Implications for Society

These trends carry profound implications for society. The democratization of AI, although fostering innovation, could exacerbate inequality if access to AI tools is uneven. Generative AI's creative potential raises excitement and concerns about misuse, such as creating deep fakes. Ethical and explainable AI is crucial for maintaining public trust, mainly as AI systems handle increasingly sensitive data. Augmented working could enhance efficiency and lead to job displacement if not managed mindfully. Finally, the push for sustainable AI mirrors society's broader demand for environmentally responsible practices.

Emerging AI technologies are pushing the boundaries of what is possible. New fields, like neuro symbolic AI, which combines neural networks with symbolic reasoning, and quantum machine learning, which leverages the principles of quantum physics to enhance machine learning algorithms, promise to create new possibilities and dramatically increase the capabilities of AI.

Ethics and Regulation

The rapid evolution of AI technologies has brought into focus numerous ethical considerations. The implications of AI extend beyond technical advancements, touching on critical areas such as societal norms, individual rights, and global regulatory compliance.

The European Union (EU) has addressed these concerns, providing a model for other regions and organizations worldwide. Recognizing the potential for AI to impact fundamental human rights, the EU has proposed an extensive regulatory framework designed to safeguard these rights in an increasingly AI-centric world ("Artificial Intelligence Act," 2022).

This proposed regulation, grounded in transparency, accountability, and data privacy, seeks to specifically regulate high-risk AI applications that could infringe upon fundamental rights. These high-risk areas might include applications in healthcare, law enforcement, critical infrastructure, and others where the AI system's decisions could have significant repercussions.

Under the proposed regulations, AI developers and users are expected to meet several requirements. These include ensuring their systems are transparent in their operations, making it clear how decisions are made, and being accountable for any mistakes or misuse.

Furthermore, data privacy is emphasized, requiring robust data management and protection measures to be in place, minimizing the risk of data breaches and misuse of personal information. This regulatory focus aligns with the EU's General Data Protection Regulation (GDPR), which sets strict data protection and privacy standards.

The regulation also calls for an AI "watchdog" or regulator that will oversee compliance, handle reports of violations, and have the authority to impose penalties for non-compliance. This serves as a deterrent for misuse of AI and fosters a culture of ethical AI practices.

While the EU's proposed AI regulation provides a comprehensive framework, it is essential to note that AI regulation must be dynamic, evolving alongside the technology it seeks to govern. Revisiting and refining regulatory frameworks will be critical as we continue to discover and unlock new AI capabilities.

AI ethics and regulation is not a stand-alone challenge for policy-makers and legal experts; it is a collective responsibility. Business and technology leaders have a pivotal role in supporting these regulations, encouraging ethical use of AI in their organizations, and actively participating in the ongoing dialogue about balancing AI innovation with societal wellbeing.

As we look to the future, AI regulation will play a central role in managing the impact of AI technologies, ensuring they are used responsibly and contribute positively to society. The EU's proposed regulation is an important step, and leaders must understand its implications and align their AI strategies accordingly.

The AI and Human Symbiosis: Future Scenarios

Scenario 1: AI Dominates, Humans Adapt

In this scenario, we can envision a future where AI dominates various sectors, from creative arts to critical business decision-making. With advancements in generative AI, machines can create content that rivals the quality of human-created content in many respects. Consequently, humans must adapt to this new world, where their roles shift from creators to curators, critics, or even consumers of AI-generated content.

This might sound disheartening, but it could also be seen as a new era of human creativity. With AI taking over routine and predictable tasks, humans can engage in higher-order creative thinking, emotional work, and relationship building. This could lead to a renaissance, where humans rediscover and reassert their unique qualities that machines cannot replicate.

However, this scenario also comes with challenges. As AI-generated content becomes more ubiquitous, it will be harder to discern between AI-generated and human-generated content. The authenticity of content could become a significant concern, leading to the emergence of new technologies or services that validate content authenticity. Moreover, businesses and regulatory bodies must ensure AI's ethical and responsible use, considering copyright, privacy, and content regulation issues.

Scenario 2: AI and Humans Co-Create

Another future scenario envisions AI and humans working hand in hand to co-create content, products, or services. In this scenario, AI does not replace humans but enhances human capabilities. AI serves as a tool that can amplify human creativity, allowing humans to achieve more than they could on their own.

For instance, AI could generate initial ideas or drafts, which humans then refine and improve. Alternatively, AI could handle large volumes of data or repetitive tasks, freeing humans to focus on more strategic or creative aspects of work. The co-creation process could lead to outputs that neither humans nor AI could produce independently, combining the efficiency and scalability of AI with the creativity and emotional depth of humans.

From a business perspective, this scenario offers exciting innovation and value-creation opportunities. Businesses can develop new products, services, or content that leverage the strengths of both humans and AI. However, this approach also requires careful thought about how to integrate AI into human workflows best and how to manage the potential challenges that might arise, such as job displacement or skill mismatches.

Scenario 3 : Humans Retain Dominance, AI Supports

The third scenario imagines a future where humans dominate creative and decision-making roles, and AI primarily supports human activities. AI is seen as an assistant rather than a replacement or co-creator in this scenario. AI technologies are used to automate routine tasks, provide insights from data, or enhance the efficiency of human work, but the final decision or creative output remains in human hands.

This scenario underscores the enduring value of human judgment, creativity, and empathy. It suggests that while AI can augment human capabilities, it cannot replace the unique qualities that humans bring to the table (Sharma, 2023). This could lead to a future where human skills are even more highly valued, and the authenticity premium for human-created content or services becomes even more pronounced.

In this world, businesses must balance leveraging AI capabilities and nurturing human skills. They will need to carefully consider integrating AI into their operations in a way that enhances, rather than undermines, human work. They will also need to address the ethical and social implications of AI use, ensuring that it is used responsibly, and its benefits are distributed fairly.

Democratization of AI

The democratization of AI represents a significant shift in the field of technology, allowing a broad array of individuals and organizations to leverage AI's transformative power. By breaking down barriers of access and affordability, democratization enables AI to be used beyond the confines of tech giants, research labs, and affluent organizations, leading to its widespread application across various sectors (Thiel, 2022).

Concept of Democratization

In the context of technology, the term "democratization" signifies the process where access to a particular technology expands from a concentrated group of experts or organizations to a broader audience (Shukla, 2019). It facilitates technology's widespread availability and usability, regardless of a user's technical expertise. This process is currently underway for AI, opening doors for its application across various industries and sectors.

The democratization of AI is making AI technology more accessible to non-experts and improving the usability of AI tools. This includes the availability of pre-trained models, automated machine learning platforms, and cloud-based AI services, which drive down the cost and complexity of creating AI models. As a result, businesses of all sizes, researchers, and even hobbyists can leverage AI for their specific needs (Thiel, 2022).

The accessibility of AI and ML technologies has skyrocketed due to recent groundbreaking advancements. For instance, significant progress in natural language processing—an AI technique that facilitates machine comprehension of human language—has given rise to applications like OpenAI's ChatGPT. Machine learning algorithms have progressively evolved, gaining the ability to draw from data and formulate predictions or decisions without the need for explicit programming for these functions. However, the broadened use of AI and ML is not purely a result of technological advancements.

The emergence of budget-friendly, intuitive tools has been a key contributor to this transformation. Open-source platforms such as TensorFlow and PyTorch have supplied the essential structure for constructing machine learning models. These platforms are freely available and are upheld by large communities that regularly offer improvements and updates, making them an invaluable resource for entities aiming to incorporate ML into their workflows.

Additionally, cloud-based solutions such as Google Cloud ML Engine, Amazon SageMaker, and Microsoft Azure ML have lowered the entry threshold by delivering scalable machine learning services with low setup demands. These platforms offer pre-configured models, automated machine-learning capabilities, and instruments for constructing, training, and deploying models, thereby circumventing the requirement for expensive hardware or specialized expertise.

Most recently, OpenAI, followed by Google's Bard, has made the headlines with mainstream access to AI tools. Solutions like Midjourney can now generate indistinguishable graphics from real cameras or graphics for use that would put most gig graphic artists out of business.

Business Transformation

With these capabilities becoming widely available and more affordable, businesses and individuals across the spectrum now have unparalleled access to AI and ML resources. These tools can dramatically bolster their operations, efficiency, and productivity, allowing them to tap into their proprietary and public data without the financial burden of employing a team of data scientists or the need to create high-cost bespoke models.

This level of transformation presents a business risk. Information services companies, for example, are on the path of being disintermediated by small players. While big players, like Amazon, were seen in the past as potential market entrants and disruptors, AI's democratization significantly lowers the entry barrier.

The differentiator for companies will be the private wealth of data they aggregate, develop, and enhance over time. The power of "private intelligence" enables corporations to extract vital insights, modifying models to fit their specific needs and situations. This aspect has also become particularly crucial in light of the growing importance of data privacy and security. With the introduction of laws like the General Data Protection Regulation (GDPR) in Europe and the California Consumer Privacy Act (CCPA) in the U.S., businesses are mandated to manage personal data responsibly. Training models on private data allows them to exploit the benefits of AI and ML while complying with these regulations.

In the rapidly evolving democratization of AI, the defining parameters of success have shifted dramatically. More is needed for businesses to focus solely on traditional metrics such as cost-effectiveness, delivery to market, and value addition. Instead, they must demonstrate a nimble ability to innovate swiftly, scale effectively, and leverage AI tools.

By employing affordable tools and training models on their data, firms can leverage the potency of these technologies without a substantial investment in data scientists and expensive custom models. This change could level the competitive playing field, allowing businesses of all scales to compete fairly in the increasingly data-centric economy.

However, it is manageable for businesses. The broadening access to AI and ML is poised to revolutionize businesses across various industries. For instance, in healthcare, machine learning can predict patient outcomes, optimize treatment plans, and simplify administrative tasks. Retail can utilize AI to amplify customer experiences through personalized recommendations, and manufacturing can improve efficiency via predictive maintenance and process automation.

Rise of Solopreneurs

The proliferation of affordable AI technologies has ignited an era where solopreneurs - individual entrepreneurs equipped with the power of AI and cloud tools - are emerging as a potent force. This new wave of start-ups, characterized by their agility and tech-savviness, reshapes industries, and challenges the status quo.

The competitive edge solopreneurs enjoy stems from their nimbleness and flexibility, significantly amplified by AI and cloud technologies. By harnessing these accessible tools, solopreneurs can experiment, innovate, and pivot rapidly, adapting to market trends and consumer preferences with a speed that often outpaces larger, more established organizations. The nature of their operations, typically lean and customer-focused, allows them to implement AI-powered solutions seamlessly, enhancing their offerings and streamlining their operations.

Furthermore, solopreneurs increasingly leverage AI to establish a footing in areas previously monopolized by larger corporations. They can utilize AI to conduct complex data analysis, make accurate predictions, automate mundane tasks, and even provide personalized customer experiences. For instance, a solopreneur running an online store can use AI tools to analyze customer behavior, predict future trends, and tailor product recommendations for each individual visitor. This level of personalization, quick decision-making, and the ability to act on insights create a compelling value proposition that can challenge even the most entrenched competitors.

AI democratization is leveling the playing field, enabling solopreneurs to punch above their weight and compete fiercely in the global marketplace.

Future of Democratized AI

Looking ahead, the democratization of AI is expected to continue, powered by advancements in AI technology and growing societal and business interest in leveraging AI. As AI becomes increasingly integrated into everyday technologies, we can expect a more profound impact on society, including changes to how we work, communicate, and interact with the world around us (Thiel, 2022). As the democratization of AI progresses, it is crucial to navigate the opportunities and challenges it presents. By fostering an environment that promotes the responsible and ethical use of AI while making it more accessible, we can harness the power of AI to create a better future.

Advice for Leaders

Leaders must have a broad understanding of AI, even if they are not fully immersed in the technical details.

Business and tech leaders need to recognize that integrating AI into their organizations is not just about adopting new technology—it is also about fostering a culture of innovation, investing in talent growth, and forming strategic partnerships. Understanding AI is not just about learning how to code—it is about understanding its potential and limitations, ethical implications, and impact on society.

Business and technology leaders should be prepared to adapt their strategies as AI evolves. They should be ready to invest in new technologies, retrain their employees, and rethink their business models in light of these changes.

Leaders should also be proactive about addressing the ethical implications of AI. They must ensure that their AI systems are transparent, accountable, and unbiased. They also need to consider how their use of AI could impact their customers, their employees, and society.

Ultimately, the most successful leaders will be those who embrace AI's opportunities, are willing to adapt and innovate, and are committed to using AI in a way that benefits all.

Path Forward

As we move forward, we can anticipate a future where AI is a pervasive part of our lives. It will become more integrated into our daily routines and play a significant role in various sectors, from healthcare to education, transportation, and beyond.

AI will continue to evolve, bringing new capabilities and challenges that will dramatically improve how we live and work.

However, business and technology leaders must monitor and actively engage with these developments. By doing so, they can ensure that they are prepared for the changes AI will bring and can effectively use these new tools to their advantage.

Ultimately, the future of AI will be shaped by how we choose to use it. It is up to us to harness its potential for the betterment of society, ensuring that it is used in a way that benefits all, not just a select few.

ACKNOWLEDGMENTS

The journey of writing is a mirror to the soul. As I journeyed through the creation of this book, I recalled the individuals who had faith in me, took risks in betting on me, and offered unwavering support. Before acknowledging any individuals, my first and foremost gratitude is extended to God for all His ceaseless blessings and strengthening trials. The answers to my prayers have held me strong, and His guidance is the bedrock upon which I find meaning in life.

Upon my arrival in the United States 25 years ago as an international student, Raj Murphy, Michelle Norwood, and Michele Schweitz saw potential in an 18-year-old and gave me an opportunity in a computer lab, as well as the task of building a contemporary website for the Graduate School and Research at IUP. My Uncle Abbas was my family away from home and helped me navigate life in a new country. Dr. Giorgio Ingargiola from Temple University, despite only knowing of me from my father, generously mailed me a surprise $2,000 check to buy my first computer. I am also thankful to my professors, Jim Wolfe (IUP), Rick Adkins (IUP), and Axel Schreiner (RIT). Their engaging approach and ground-breaking methods of research and innovation were a true inspiration.

In college, I encountered a fellow Iraqi who nudged me towards website development for Iraqi soccer fans, introducing me to the then-new .NET framework from Microsoft. Though the connection did not last, the experience proved invaluable and life-changing, leading me to a position at Johnson & Johnson (J&J) shortly after, building innovative intranets and advanced search engines empowering several J&J websites.

There, under the mentorship of Joe Napoli, I learned to lead with empathy and put customers first. I owe the kickstart of my career with IHS to Mike Craig, who championed my career there. Despite his senior position, he was generous with his time, wisdom, and coaching and continuously presented me with opportunities to excel and demonstrate my abilities. My managers Anthony Lopez, Paul Balas, Dan Martel, Keith Worfolk, and Rich Herrmann were great mentors. Kim Brown was an example of a savvy technical leader who is business-aligned, and her coaching and support were tremendous. Jo Moon, reporting to the CEO then, saw something in me that she brought me on to her leadership team of senior directors and VPs even as I was just an associate director at the time. Over five years, her wisdom and insights were invaluable in helping me leap to more senior positions and develop my executive presence. I am fortunate to have worked with leaders such as Scott Key, Jerre Stead, John Oechsle, Rich Walker, Stephanie Buscemi (who later became CMO for Salesforce and Confluence), and Sari Granat, who were phenomenal leaders and unique in their domains and very supportive of my professional journey. I am grateful to Jamey Rosenfield, a pioneer in energy and innovation. He gave me his trust and support in overhauling the end-to-end digital experience for the flagship CERAWeek event.

Later, at IHS Markit, Chad Moss, the Global CIO then, asked me to join his technology leadership team, where I served with him for over six years. He is one of the most genuine, intelligent, kind, insightful, and articulate people I have ever worked with. Over the years, he challenged and coached me and was the first to recognize my successes. He has impacted my career so much that I am forever indebted to him. My peers in leadership—especially Paul, Ian, and Leah—were all outstanding and I learned a great deal from all of them. I would also like to recognize my esteemed faculty at the Executive MBA program at the University of Denver, especially Kerry Plemmons, who has further inspired my innovation endeavors. The 18-month program was transformational and prepared me fully to be the executive I am today.

The inception of writing this book was spurred by Yaacov Mutnikas, Chief Data Scientist and later CTO of Market Intelligence at S&P Global. His challenge to present a solution in seven days for operational difficulties leaders face set me on the path of machine learning research and eventually writing this book.

Lastly, my gratitude goes to my parents for their love and support. They gave me their life savings during difficult times so I could pursue my dream of studying in the US. My father's inspiring life trajectory from a modest background to a tenured professor at an Ivy League, with esteemed international awards from kings and heads of state, worldwide recognition, and more than 40 published books in Arabic and English, and my mother's selflessness and love for her family despite her esteemed academic role for four decades, have always served as my guiding principles in life. My sisters' unconditional love has always reminded me to love others and treat them kindly. My wife, Hawra, with her golden heart, kindness, and positivity, supported me through the lows and highs of life with her beautiful smile. I thank my children, Malaak, Ali, and Muhsin, who were patient and understanding while I completed this project. They bring me joy every moment of my day, and I am fortunate to have them.

REFERENCES

Author's Notes

Agrawal, A., Gans, J., & Goldfarb, A. (2022, December 14). *Chatgpt and how AI disrupts industries.* Harvard Business Review. https://hbr.org/2022/12/chatgpt-and-how-ai-disrupts-industries

Crevier, D. (1993). The tumultuous history of the search for Artificial Intelligence. Basic Books.

McCorduck, P. (2019). Machines who think: A personal inquiry into the history and Prospects of Artificial Intelligence. CRC Press.

Nilsson, N. J. (2010). The quest for artificial intelligence: A history of ideas and achievements. Cambridge University Press.

Chapter 1 - History

Abbass, H. (2022, September 15). *An AI professor explains: Three concerns about granting citizenship to Robot Sophia.* The Conversation. https://theconversation.com/an-ai-professor-explains-three-concerns-about-granting-citizenship-to-robot-sophia-86479

Agrawal, A., Gans, A., & Goldfarb, A. (2022, December 14). *Chatgpt and how AI disrupts industries.* Harvard Business Review. https://hbr.org/2022/12/chatgpt-and-how-ai-disrupts-industries

Ai Art Generator – adobe firefly. (n.d.). https://www.adobe.com/sensei/generative-ai/firefly.html

Ai trends: What's shaping our future. XWYRD Blog. (2023, April 29). https://www.xwyrd.com/ai-trends-whats-shaping-our-future/

Barazy, M. (2023, May 8). AI: Cognitive Technology as a new world view and paradigm ... - linkedin. https://www.linkedin.com/pulse/ai-cognitive-technology-new-world-view-paradigm-shift-maan-barazy

Brin, S., & Page, L. (1998). The anatomy of a large-scale hypertextual web search engine. *Computer Networks and ISDN Systems, 30*(1–7), 107–117. https://doi.org/10.1016/s0169-7552(98)00110-x

Bron, D. (n.d.). *Living in a post-scarcity society: How automation, AI, and universal basic income could reshape the global economy.* LinkedIn. https://www.linkedin.com/pulse/living-post-scarcity-society-how-automation-ai-basic-income-daniel/

Crevier, D. (1993). The tumultuous history of the search for Artificial Intelligence. Basic Books.

Frąckiewicz, M. (2023, May 25). *The advent of AI: A complete guide to building intelligent systems.* TS2 SPACE. https://ts2.space/en/the-advent-of-ai-a-complete-guide-to-building-intelligent-systems/

Gartner Hype Cycle Research Methodology. Gartner. (n.d.). https://www.gartner.com/en/research/methodologies/gartner-hype-cycle

Gold, E. (2023, April 10). The history of artificial intelligence from the 1950s to today. https://www.freecodecamp.org/news/the-history-of-ai/

Henderson, H. (2007). Artificial Intelligence: Mirrors for the mind. Chelsea House.

IBM. (2021). *What is machine learning?*. IBM. https://www.ibm.com/topics/machine-learning

Ijarotimi, T. (2023, March 29). *Chatgpt and its potential for job replacement: A comprehensive analysis.* Interesting Engineering. https://interestingengineering.com/innovation/chatgpt-potential-job-replacement-analysis

Johnson, A. (2023, May 23). *I turned my vacation photos into nightmares with Photoshop's new Generative AI Tool.* The Verge. https://www.theverge.com/2023/5/23/23734821/photoshop-generative-ai-fill-tool-adobe

Kumar, R. (2023, May 28). *A beginner's Guide to Artificial Intelligence and machine learning.* BigDev. https://bigdev.space/beginners-guide-to-artificial-intelligence-and-machine-learning/

Li, N. (2023, May 1). *Another AI-generated Drake track is going viral.* Hypebeast. https://hypebeast.com/2023/4/drake-ai-generated-not-a-game-track-listen-info

McCorduck, P. (2019). Machines who think: A personal inquiry into the history and Prospects of Artificial Intelligence. CRC Press.

Next Tech AI. (2023). *Bill Gates and Socrates talking about AI.* Retrieved May 31, 2023, from https://www.youtube.com/watch?v=hJ5qN9PRmFc.

Nilsson, N. J. (2010). The quest for artificial intelligence: A history of ideas and achievements. Cambridge University Press.

Oreck, A. (n.d.). *Modern jewish history.* The Golem. https://www.jewishvirtuallibrary.org/the-golem

Pai, A. (2020, May 26). *What is tokenization in NLP? here's All you need to know.* Analytics Vidhya. https://www.analyticsvidhya.com/blog/2020/05/what-is-tokenization-nlp/

Research Djve. (n.d.). Global AI Accelerator Chip Market Analysis. https://www.researchdive.com/8599/ai-accelerator-chip-market

Riskin, J. (2018). The restless clock: A history of the centuries-long argument over what makes Living things tick. The University of Chicago Press.

Russell, S. J. (2016). Artificial Intelligence: A modern approach. Pearson.

Shashkevich-Stanford, A. (2019, March 5). *Greek myths have some scary ideas about robots and A.I.* Futurity. https://www.futurity.org/artificial-intelligence-greek-myths-1999792/

Shroff, G. (2020). The intelligent web: Search, smart algorithms, and Big Data. Oxford University Press.

Smith, C. (2006, December). The history of Artificial Intelligence - University of Washington. https://courses.cs.washington.edu/courses/csep590/06au/projects/history-ai.pdf

Turing, A. M., & Copeland, B. J. (2013). The essential turing: Seminal writings in Computing, Logic, philosophy, artificial intelligence, and artificial life "Plus" the secrets of enigma. Oxford University Press.

Wiles, J., & Jaffri, A. (n.d.). *What's new in Artificial Intelligence from the 2022 Gartner Hype cycle.* Gartner. https://www.gartner.com/en/articles/what-s-new-in-artificial-intelligence-from-the-2022-gartner-hype-cycle

Yıldız, M. (2023, April 13). *What is machine learning? introduction to ML - Clarusway.* Online IT Bootcamp; Learn Coding, Data Science, AWS, DevOps, Cyber Security & Salesforce. https://clarusway.com/what-is-machine-learning

Chapter 2 – Reshaping Business

Al-Ghourabi, A. (n.d.). The democratization of AI: The rise of solopreneurs and the Evolution of Business. https://www.linkedin.com/pulse/democratization-ai-rise-solopreneurs-evolution-ad-al-ghourabi

ALONSO, S. (2023, April 12). *Women are diagnosed years later than men for same diseases, Danish study*... Soledad ALONSO on LinkedIn: Women are diagnosed years later than men for same diseases, Danish study... https://www.linkedin.com/posts/soledadalonso_women-are-diagnosed-years-later-than-men-activity-7051891636972183552-R15H/?originalSubdomain=cz

Coward, R. (2023, January 19). *Is ai going to take our jobs?*. Is AI going to take our jobs? https://richcoward.com/is-ai-going-to-take-our-jobs/

Davenport, T. H., & Ronanki, R. (2022, November 7). *3 things AI can already do for your company*. Harvard Business Review. https://hbr.org/2018/01/artificial-intelligence-for-the-real-world

Dawood, M. (2023, April 10). *Dr Muhammad Dawood on linkedin: #cybersecurity #cyber #penetrationtesting #cyberattack*... Dr Muhammad Dawood on LinkedIn: #cybersecurity #cyber #penetrationtesting #cyberattack... https://www.linkedin.com/posts/muhammad-dawoud_cybersecurity-cyber-penetrationtesting-activity-7051118013839740928-DZJV/?originalSubdomain=gh

Digital Technology Guru. (2023, June 7). *The promising future of AI-enabled web development*. Digital Technology Guru. https://dtgreviews.com/uncategorised/the-promising-future-of-ai-enabled-web-development/38671/

Exploring how AI transforms sales processes. Dataconomy. (2023, May 17). https://dataconomy.com/2023/05/17/artificial-intelligence-in-sales-101/

Frąckiewicz, M. (2023a, May 11). *The role of Artificial Intelligence in Industrial Wireless Sensor Networks*. TS2 SPACE. https://ts2.space/en/the-role-of-artificial-intelligence-in-industrial-wireless-sensor-networks/

Frąckiewicz, M. (2023b, May 18). *How ai is revolutionizing the app development industry: Opportunities for growth and Profit*. TS2 SPACE. https://ts2.space/en/how-ai-is-revolutionizing-the-app-development-industry-opportunities-for-growth-and-profit/

Ghraichy, J. (2023, April 6). *The power of personalization: Targeting your audience in digital advertising*. ▪. https://www.jghraichy.com/blog/the-power-of-personalization-targeting-your-audience-in-digital-advertising

The Great AI Debate: A Chat Between ChatGPT and Dragonfly. AI Talks. (2023, April 1). https://ai-talks.org/2023/03/31/the-great-ai-debate-a-chat-between-chatgpt-and-dragonfly/

Gudigar, A., Raghavendra, U., Nayak, S., Ooi, C. P., Chan, W. Y., Gangavarapu, M. R., Dharmik, C., Samanth, J., Kadri, N. A., Hasikin, K., Barua, P. D., Chakraborty, S., Ciaccio, E. J., & Acharya, U. R. (2021). Role of artificial intelligence in COVID-19 detection. *Sensors, 21*(23), 8045. https://doi.org/10.3390/s21238045

Hopwood, K. (2022, May). *CarMax puts customers first with car research tools powered by Azure OpenAI Service*. Microsoft Customers Stories. https://customers.microsoft.com/en-us/story/1501304071775762777-carmax-retailer-azure-openai-service

Insighteurs. (2023, March 20). *How artificial intelligence is revolutionizing the customer experience*. Insighteurs. https://www.insighteurs.com/how-artificial-intelligence-is-revolutionizing-the-customer-experience/

Jacob, K. (2021, February). A Chicago Fintech Financial Health Collaboration: Springfour and Enova. https://springfourdirect.com/document/Enova_CaseStudy_2021.pdf

Joseph. (2023, April 30). *How ai will disrupt the mobile app design and Development Industry: A simple guide*. MakeThatContent. https://makethatcontent.com/how-ai-will-disrupt-the-mobile-app-design-and-development-industry-a-simple-guide/

Karandish, D. (2021, June). *7 benefits of AI in Education*. THE Journal. https://thejournal.com/articles/2021/06/23/7-benefits-of-ai-in-education.aspx

Keyser, J. (2023, February 13). *Cybersecurity: The power of combining soar and Siem*. Aliado Solutions. https://aliadosolutions.com/cybersecurity-the-power-of-combining-soar-and-siem/

Koopman, L. (n.d.). *6 major impacts of artificial intelligence*. LinkedIn. https://www.linkedin.com/pulse/6-major-impacts-artificial-intelligence-logan-koopman/

Meah, J. (2023, April 26). *The Digital Revolution in banking: Exploring the future of Finance*. Techopedia. https://www.techopedia.com/the-digital-revolution-in-banking-exploring-the-future-of-finance

Mustapha, A. (2023, June). *Artificial Intelligence*. Education. https://vocal.media/education/artificial-intelligence-z7gq05vp

Napilay, J. (n.d.). *How social media and Seo work together to boost online success*. LinkedIn. https://www.linkedin.com/pulse/how-social-media-seo-work-together-boost-online-success-napilay/

Pandey, K., & Pandey, A. K. (2022, May 13). *The power of introverts at the Workplace*. Jumpstart Magazine. https://www.jumpstartmag.com/the-power-of-introverts-at-the-workplace/

The power of Ai Chatbots: How CHATGPT is Revolutionizing Customer Service. Otter.ai. (n.d.). https://otter.ai/blog/the-power-of-ai-chatbots-how-chatgpt-is-revolutionizing-customer-service

Raviv, I. (2021, June). *4 ways AI can help us enter a new age of Cybersecurity*. World Economic Forum. https://www.weforum.org/agenda/2021/06/4-ways-ai-new-age-of-cybersecurity/

S&P Global. (2023). *https://investorfactbook.spglobal.com/sp-global/kensho-a-hub-for-innovation-and-transformation/*. Kensho Technologies. https://investorfactbook.spglobal.com/sp-global/kensho-a-hub-for-innovation-and-transformation/

Spataro, J. (2023, May 16). *Introducing Microsoft 365 copilot – your copilot for work*. The Official Microsoft Blog. https://blogs.microsoft.com/blog/2023/03/16/introducing-microsoft-365-copilot-your-copilot-for-work/

Spizheva, D. (2023, May 31). *Ai examples in business: How do companies use artificial intelligence?: Turnkey AI*. TurnKey Labs. https://turnkey-labs.com/tech-trends/businesses-using-ai/

Switzerland, F. (2016, June 23). *Top Fintech startups over last decades*. Fintech Schweiz Digital Finance News - FintechNewsCH. http://fintechnews.ch/fintech/top-fintech-startups-over-last-decades/3160/

Unleashing the power of enterprise AI: Transforming decision-making Beyond Data Science. Learnow. (n.d.). https://www.learnow.live/blog/enterprise-ai-is-moving-beyond-data-science-and-into-every-decision

Wiggers, K. (2022, May 24). *Microsoft expands Azure openai service with new features*. TechCrunch. https://techcrunch.com/2022/05/24/microsoft-expands-azure-openai-service-with-fine-tuning-features-and-more/

Wilde, S., & Guthrie, P. (2023, May 15). *Scale up archives*. NVOY Technologies. https://nvoytechnologies.com/category/scale-up

Chapter 3 – Building an AI Culture

Ammanath, B. (2022, June 29). *Ai transformation and culture shifts*. Deloitte United States. https://www2.deloitte.com/us/en/pages/technology/articles/build-ai-ready-culture.html

Bruhn, C. (2022, May 10). *Council Post: Strengthening Company culture with Artificial Intelligence*. Forbes. https://www.forbes.com/sites/forbesbusinesscouncil/2022/05/09/strengthening-company-culture-with-artificial-intelligence/

English, L. (2023, May 30). *The impact of AI on company culture and how to prepare now*. Forbes. https://www.forbes.com/sites/larryenglish/2023/05/25/the-impact-of-ai-on-company-culture-and-how-to-prepare-now/

Haefner, N., Wincent, J., Parida, V., & Gassmann, O. (2020, October 18). *Artificial Intelligence and Innovation Management: A review, Framework, and research agenda☆*. Technological Forecasting and Social Change. https://www.sciencedirect.com/science/article/pii/S004016252031218X

Lee, B., Morgan, K., & Choo, E. (2020, August 27). *Scaling AI value with Agile Ai*. Accenture. https://www.accenture.com/us-en/insights/applied-intelligence/scale-ai-agile

Morace, C. (2014). Transform: How leading companies are winning with disruptive social technology. McGraw-Hill Education.

Neeley , T., & Leonardi, P. (2022, April 12). *Developing a digital mindset*. Harvard Business Review. https://hbr.org/2022/05/developing-a-digital-mindset

Chapter 4 – Building AI Strategy and Team

D'Silva, V., & Lawler, B. (2022, February 28). *What makes a company successful at using AI?*. Harvard Business Review. https://hbr.org/2022/02/what-makes-a-company-successful-at-using-ai

Fountaine, T., McCarthy, B., & Saleh, T. (2020, June 1). *Building the AI-powered organization*. Harvard Business Review. https://hbr.org/2019/07/building-the-ai-powered-organization

Goasduff, L. (2019, September 18). *3 barriers to AI adoption*. Gartner. https://www.gartner.com/smarterwithgartner/3-barriers-to-ai-adoption

Gurtu, A. (2021, July 28). *Council post: How to build a perfect AI team.* Forbes. https://www.forbes.com/sites/forbestechcouncil/2021/07/28/how-to-build-a-perfect-ai-team/

Ng, A. (2019a, June 5). *Forbes Insights: How to build a great AI team.* Forbes. https://www.forbes.com/sites/insights-intelai/2019/05/22/how-to-build-a-great-ai-team/

Ng, A. (2019b, August 21). *Andrew Ng: How to choose your first AI project.* Harvard Business Review. https://hbr.org/2019/02/how-to-choose-your-first-ai-project

Nguyen, H. (2022, November 7). How to build your own AI software with an in-house AI team. https://www.orientsoftware.com/blog/how-to-build-ai-software/

Mohan, S. (2022, May 30). *Council post: How to make it easier to implement AI in your business.* Forbes. https://www.forbes.com/sites/forbestechcouncil/2022/05/27/how-to-make-it-easier-to-implement-ai-in-your-business/

Vartak, M. (2022, November 10). *How to scale AI in your organization.* Harvard Business Review. https://hbr.org/2022/03/how-to-scale-ai-in-your-organization

Walch, K., & Schmelzer, R. (2021, April 6). *How to build a machine learning model in 7 steps: TechTarget.* Enterprise AI. https://www.techtarget.com/searchenterpriseai/feature/How-to-build-a-machine-learning-model-in-7-steps

Walden, S. (2022, May 2). *Overcoming barriers to AI adoption: Experts weigh in.* Dell. https://www.dell.com/en-us/perspectives/overcoming-barriers-to-ai-adoption-experts-weigh-in/

Williams, B., & Mattar, M. (2022, November 1). *Training and building machine learning models: Scale Ai.* ScaleAI. https://scale.com/guides/model-training-building

Chapter 5 – Ethics and Privacy

24 artificial intelligence examples that are changing the world. InsideAIML. (n.d.). https://insideaiml.com/blog/24-artificial-intelligence-examples-that-are-changing-the-world-1243

Ai Ethics in action. IBM. (n.d.). https://www.ibm.com/thought-leadership/institute-business-value/en-us/report/ai-ethics-in-action

Ai in people analytics and HR digitalisation: Privacy, transparency and trust issues: People analytics world 2023. (n.d.). https://www.peopleanalyticsworld.com/2023/ai-people-analytics-hr-digitalisation-privacy-transparency-trust

Banta, N. (2023, May). What are the ethical implications of advancements in ai?. B12. https://www.b12.io/resource-center/ai-thought-leadership/what-are-the-ethical-implications-of-advancements-in-ai.html

Bias in AI: What biases do marketers need to be careful about. MarTech Series. (2023, February 28). https://martechseries.com/mts-insights/staff-writers/bias-in-ai-what-biases-do-marketers-advertisers-need-to-be-careful-about/

C., S. (n.d.). *Bias in bots: Understanding the impact of bias in AI.* LinkedIn. https://www.linkedin.com/pulse/bias-bots-understanding-impact-ai-selina-castorena/

EU General Data Protection Regulation (GDPR): An implementation and compliance guide. (2020). IT Governance Publishing.

Frąckiewicz, M. (2023, May 13). *The importance of transparency in Artificial Intelligence.* TS2 SPACE. https://ts2.space/en/the-importance-of-transparency-in-artificial-intelligence/

Fugulin, M. de A. (2023, May 11). *Developing ethical services in the age of AI: Building Trust and Responsibility.* LinkedIn. https://www.linkedin.com/pulse/developing-ethical-services-age-ai-building-trust-de-almeida-fugulin/

G., I. (2023, May). *Building AI ethics: Safeguarding AI systems from "self" and "human interference."* LinkedIn. https://www.linkedin.com/pulse/building-ai-ethics-safeguarding-systems-from-self-human-indy-g/

Healey, R. (2023, May 1). *Why privacy by design is essential for AI projects.* Formiti. https://formiti.com/why-privacy-by-design-is-essential-for-ai-projects/

Kerry, C. F. (2022, March 9). *Protecting privacy in an AI-Driven World.* Brookings. https://www.brookings.edu/research/protecting-privacy-in-an-ai-driven-world/

Khan, S. (2023, April 30). *Chapter 5 - Healthcare and medicine: Ai's impact on diagnosis, treatment, and prevention.* issuu. https://issuu.com/khansalik53/docs/beyond_human_the_rise_of_artificial_intelligen ce_/s/23566854

Li, L., Zeng, Z., Zhang, G., Duan, K., Liu, B., & Cai, X. (2022, September 4). Exploring the individualized effect of climatic drivers on Modis net primary productivity through an explainable machine learning framework. MDPI. https://www.mdpi.com/2072-4292/14/17/4401/htm

May, A. (2023, June 8). *Best practices for protecting sensitive health information.* Wisterm. https://wisterm.com/employee-benefits/employee-benefits-compliance/best-practices-for-protecting-sensitive-health-information/

Microsoft AI. (2021). Putting principles into practice: How we approach responsible AI at Microsoft. https://www.microsoft.com/cms/api/am/binary/RE4pKH5

Yaqoob, T. (2023, April 18). *Ethical considerations in AI development and deployment.* Cointelegraph. https://cointelegraph.com/explained/ethical-considerations-in-ai-development-and-deployment

Chapter 6 - Executing Your AI Strategy

Al Bannai, F. (2023, May 25). UAE's Technology Innovation Institute launches open-source "Falcon 40B" Large language model for research & commercial utilization. Technology Innovation Institute. https://www.tii.ae/news/uaes-technology-innovation-institute-launches-open-source-falcon-40b-large-language-model

Antosz, D. (2021, July 16). The role of Change Management when implementing AI. Salesforce. https://www.salesforce.com/ca/blog/2021/07/the-role-of-change-management-when-implementing-ai-.html

Bergeret, B., & Malaurent, J. (2023, March). GETTING COMPANIES ON BOARD WITH RESPONSIBLE ARTIFICIAL INTELLIGENCE. https://amchamfrance.org/wp-content/uploads/2023/03/AI-White-Paper-14-Mars-2022-2_compressed.pdf

Candelon, F., Charme di Carlo, R., Bondt, M. D., & Evgeniou, T. (2021). Ai regulation is coming. Harvard Business Review. https://hbr.org/2021/09/ai-regulation-is-coming

Cath, C. (2018). Governing Artificial Intelligence: Ethical, legal and technical opportunities and challenges. Philosophical Transactions of the Royal Society A: Mathematical, Physical

and Engineering Sciences, 376(2133), 20180080.
https://doi.org/10.1098/rsta.2018.0080

Cockburn, I. M., Henderson, R., & Stern, S. (2018, March). The impact of Artificial Intelligence on Innovation - NBER. https://www.nber.org/system/files/working_papers/w24449/w24449.pdf

Cohen, M. (2023, April 25). Chatgpt and Ai are the next worker recruitment and Retention Game Changers. CNBC. https://www.cnbc.com/2023/04/25/chatgpt-and-ai-are-talent-recruitment-and-retention-game-changers.html

Collins, C. (2022, July 6). What is the difference between the data acquisition and data exploration in an AI project cycle? - 2023 guide. Revenues & Profits. https://revenuesandprofits.com/difference-between-data-acquisition-exploration-ai-project-cycle/

Davenport, T. H., & Redman, T. C. (2022, December 20). How AI is Improving Data Management. MIT Sloan Management Review. https://sloanreview.mit.edu/article/how-ai-is-improving-data-management/

Davenport, T. H., & Ronanki, R. (2022, November 7). 3 things AI can already do for your company. Harvard Business Review. https://hbr.org/2018/01/artificial-intelligence-for-the-real-world

Dawson, J. (2020, July 3). The rise of AI in Talent Acquisition. Ideal. https://ideal.com/the-rise-of-ai-in-talent-acquisition/

Ethics and governance of AI. Ethics and Governance of AI | Berkman Klein Center. (n.d.). https://cyber.harvard.edu/topics/ethics-and-governance-ai

The fox MIS professional achievement program. Professional Achievement Program. (n.d.). https://community.mis.temple.edu/professionalachievement/calendar/action~agenda/page_offset~-1/time_limit~1572930000/cat_ids~142406,142407,2511,79070/request_format~json/

Joshi, N. (2019, July 24). Why regulatory compliance can be complicated and how AI can simplify it. Forbes. https://www.forbes.com/sites/cognitiveworld/2019/07/22/why-regulatory-compliance-can-be-complicated-and-how-ai-can-simplify-it/

Marr, B. (2022, October 12). The 4 biggest trends in big data and analytics right for 2021. Forbes. https://www.forbes.com/sites/bernardmarr/2021/02/22/the-4-biggest-trends-in-big-data-and-analytics-right-for-2021/

Marr, B. (2023, April 14). How to make Ai work in your organization. Forbes. https://www.forbes.com/sites/bernardmarr/2023/04/13/how-to-make-ai-work-in-your-organization/

McKinsey & Company. (2020, November 17). An executive's guide to ai. McKinsey & Company. https://www.mckinsey.com/business-functions/mckinsey-analytics/our-insights/an-executives-guide-to-ai

NIST. (2023, May 26). Artificial Intelligence. NIST. https://www.nist.gov/artificial-intelligence

PwC. (2022). What are our ai partnerships?. PwC. https://www.pwc.com/gx/en/issues/data-and-analytics/artificial-intelligence/partnerships-alliances.html

Sagiraju, S. (2022, April 21). Council post: Managing the data for the AI lifecycle. Forbes. https://www.forbes.com/sites/forbestechcouncil/2022/04/20/managing-the-data-for-the-ai-lifecycle/

Schick, I. (2023, June). Forging the future together: The critical role of strategic partnerships in Harnessing generative AI in patent practice. LinkedIn. https://www.linkedin.com/pulse/forging-future-together-critical-role-strategic-ai-schick-phd-esq

Smith, S. (2023, February 9). Integration of AI into business workflow and Operations. Tollanis Solutions Inc. https://tollanis.com/ai/integration-of-ai-into-business-workflow-and-operations/

team, Profit. co. (2023, May 8). A guide to deploy AI for strategy implementation. Best OKR Software by Profit.co. https://www.profit.co/blog/okr-university/a-guide-to-deploy-ai-for-strategy-implementation

Webb, H. (n.d.). How is Ai Enabling Business Innovation?. LinkedIn. https://www.linkedin.com/pulse/how-ai-enabling-business-innovation-hannah-webb/

Chapter 7 – AI and Cybersecurity

3 ways AI will change the nature of cyber attacks. World Economic Forum. (n.d.). https://www.weforum.org/agenda/2019/06/ai-is-powering-a-new-generation-of-cyberattack-its-also-our-best-defence/

Anderson, M. (2023, May 13). *The impact of AI on Cybersecurity.* IEEE Computer Society. https://www.computer.org/publications/tech-news/trends/the-impact-of-ai-on-cybersecurity/

Artificial Intelligence (AI) for Cybersecurity. IBM. (n.d.). https://www.ibm.com/security/artificial-intelligence

Darktrace. (2023). *Company overview.* Darktrace. https://darktrace.com/company

Drexel Univ. (2021, November). *Role of artificial intelligence in Cybersecurity.* College of Computing & Informatics. https://drexel.edu/cci/stories/role-of-AI-in-cybersecurity/

Driz, S. E. (2023, March 4). *Artificial Intelligence and cybersecurity: How ai is transforming the industry.* The Driz Group. https://www.drizgroup.com/driz_group_blog/artificial-intelligence-and-cybersecurity-how-ai-is-transforming-the-industry#:~:text=Advantages%20of%20AI%20in%20Cybersecurity,-The%20use%20of&text=Increased%20speed%20and%20accuracy:%20AI,caused%20by%20a%20cyber%20attack.

Elkady, S. (2023, March). *AI in 2023: Predictions and possibilities.* LinkedIn. https://www.linkedin.com/pulse/ai-2023-predictions-possibilities-sami-elkady-/

Elliott, K. (2022). *Impact of using open source software on cybersecurity.* CyberSaint Security. https://www.cybersaint.io/blog/impact-of-using-open-source-software-on-cybersecurity

Higgins-Dunn, N. (2020, December 3). *IBM uncovers global email attack on Covid Vaccine Supply Chain.* CNBC. https://www.cnbc.com/2020/12/03/ibm-uncovers-global-email-attack-on-covid-vaccine-supply-chain-.html

Imemixx on Reddit. (2023, June). *The Impact of Artificial Intelligence on the Future of Work.* Reddit. https://www.reddit.com/r/TopicTalk/comments/149vttt/the_impact_of_artificial_intelligence_on_the/

Insights, M. T. R. (2022, February 7). *Preparing for AI-enabled cyberattacks*. MIT Technology Review. https://www.technologyreview.com/2021/04/08/1021696/preparing-for-ai-enabled-cyberattacks/

Korolov, M. (2022, May 19). *Fantastic Open Source Cybersecurity tools and where to find them.* Data Center Knowledge | News and analysis for the data center industry. https://www.datacenterknowledge.com/security/fantastic-open-source-cybersecurity-tools-and-where-find-them

Microsoft. (2023). *Microsoft Defender for Cloud - CSPM & CWPP: Microsoft Azure.* CSPM & CWPP | Microsoft Azure. https://azure.microsoft.com/en-us/services/security-center/

Moisset, S. (2023, May 25). *How security analysts can use AI in Cybersecurity.* freeCodeCamp.org. https://www.freecodecamp.org/news/how-to-use-artificial-intelligence-in-cybersecurity/

Raviv, I. (2021, June 17). *4 ways AI can help us enter a new age of Cybersecurity.* World Economic Forum. https://www.weforum.org/agenda/2021/06/4-ways-ai-new-age-of-cybersecurity/

Rees, K. (2023, January 9). *Chatgpt used by Cybercriminals to write malware.* MUO. https://www.makeuseof.com/chatgpt-used-by-cybercriminals-to-write-malware/

Violino, B. (2022, September 13). *Artificial Intelligence is playing a bigger role in cybersecurity, but the bad guys may benefit the most.* CNBC. https://www.cnbc.com/2022/09/13/ai-has-bigger-role-in-cybersecurity-but-hackers-may-benefit-the-most.html

Chapter 8 – Role of Government

Aurangzeb, U. (2023, April 25). *"unpacking the Ethical Quandaries of Artificial Intelligence: Morality, privacy, and humanity in...* Medium. https://medium.com/@umaraurangzeb/unpacking-the-ethical-quandaries-of-artificial-intelligence-morality-privacy-and-humanity-in-4867ca7f6593

Conger, K., Fausset, R., & Kovaleski, S. F. (2019, May 14). *San Francisco bans facial recognition technology.* The New York Times. https://www.nytimes.com/2019/05/14/us/facial-recognition-ban-san-francisco.html

Doyle, E., & Grattirola, H. (2022, February 18). *EU Artificial Intelligence Regulation.* Lexology. https://www.lexology.com/library/detail.aspx?g=0a7d9189-9926-443f-8b37-cccd2750a354

Felz , D., Peretti, K., & Austin, A. (2023, February 14). *AI regulation in the U.S.: What's coming, and what companies need to do in 2023.* Law.com. https://www.law.com/2023/02/14/ai-regulation-in-the-u-s-whats-coming-and-what-companies-need-to-do-in-2023/

Fung, B. (2023, May 5). *Biden Administration unveils an AI plan ahead of meeting with tech CEOS | CNN business.* CNN. https://www.cnn.com/2023/05/04/tech/white-house-ai-plan/index.html

Gargione, F. (n.d.). The Advanced Communications Technology Satellite: An Insider's account of the emergence of interactive broadband services in Space. GlobalSpec. https://www.globalspec.com/reference/52595/203279/chapter-9-the-role-of-government-in-technology-development

Hon. Donelan, M. (2023). *A pro-innovation approach to AI regulation.* GOV.UK. https://www.gov.uk/government/publications/ai-regulation-a-pro-innovation-approach/white-paper

Keane, J. (2022, May 26). *China and Europe are leading the push to regulate A.I. - one of them could set the global playbook.* CNBC. https://www.cnbc.com/2022/05/26/china-and-europe-are-leading-the-push-to-regulate-ai.html

National AI strategy. Smart Nation Singapore. (n.d.). https://www.smartnation.gov.sg/initiatives/artificial-intelligence/

Swetlitz, I., & Ross, C. (2022, September 13). *IBM pitched its Watson supercomputer as a revolution in cancer care. it's nowhere close.* STAT. https://www.statnews.com/2017/09/05/watson-ibm-cancer/

Chapter 9 – Global Economy

Bughin, J., Seong, J., Manyika, J., Chui, M., & Joshi, R. (2018, September 4). *Notes from the AI Frontier: Modeling the impact of AI on the World Economy.* McKinsey & Company. https://www.mckinsey.com/featured-insights/artificial-intelligence/notes-from-the-AI-frontier-modeling-the-impact-of-ai-on-the-world-economy

Engler, A. (2022, March 9). *Can ai model economic choices?.* Brookings. https://www.brookings.edu/research/can-ai-model-economic-choices/

Furman, J. (2019, January 1). *Ai and the economy.* Harvard Kennedy School. https://www.hks.harvard.edu/publications/ai-and-economy

Gast, A. (2022, May). *Why Artificial Intelligence is vital in the race to meet the sdgs.* World Economic Forum. https://www.weforum.org/agenda/2022/05/artificial-intelligence-sustainable-development-goals/

Goldfarb, A., & Trefler, D. (2018, January 19). *Artificial Intelligence and international trade.* NBER. https://www.nber.org/books-and-chapters/economics-artificial-intelligence-agenda/artificial-intelligence-and-international-trade

Hupfer, S. (2019, December). Global Perspectives on AI | deloitte insights. https://www2.deloitte.com/us/en/insights/focus/cognitive-technologies/global-perspectives-ai-adoption.html

Marr, B. (2023, June 5). *The 15 biggest risks of Artificial Intelligence.* Forbes. https://www.forbes.com/sites/bernardmarr/2023/06/02/the-15-biggest-risks-of-artificial-intelligence/

Meltzer, J. P. (2022a, March 9). *Maximizing AI's economic, social, and trade opportunities.* Brookings. https://www.brookings.edu/blog/up-front/2019/05/13/maximizing-ais-economic-social-and-trade-opportunities/

Meltzer, J. P. (2022b, March 9). *The impact of artificial intelligence on International Trade.* Brookings. https://www.brookings.edu/research/the-impact-of-artificial-intelligence-on-international-trade/

OECD. (2021, March). Chapter 7: Ai Policy and national strategies. https://wp.oecd.ai/app/uploads/2021/03/2021-AI-Index-Report-_Chapter-7.pdf

Patel, J., Manetti, M., Mendelsohn, M., Mills, S., Felden, F., Littig, L., & Rocha, M. (2022, September 13). *Ai brings science to the art of policymaking.* BCG Global. https://www.bcg.com/publications/2021/how-artificial-intelligence-can-shape-policy-making

Qureshi, Z. (2022, March 9). *How digital transformation is driving economic change.* Brookings. https://www.brookings.edu/blog/up-front/2022/01/18/how-digital-transformation-is-driving-economic-change/

Stahl, A. (2022, November 9). *How ai will impact the future of work and life.* Forbes. https://www.forbes.com/sites/ashleystahl/2021/03/10/how-ai-will-impact-the-future-of-work-and-life/

Tarswell, E. (2018, December 6). *Building an AI world: Report on national and regional AI Strategies.* CIFAR. https://cifar.ca/cifarnews/2018/12/06/building-an-ai-world-report-on-national-and-regional-ai-strategies/

West, D. M., & Allen, J. R. (2023, May 22). *How artificial intelligence is transforming the world.* Brookings. https://www.brookings.edu/research/how-artificial-intelligence-is-transforming-the-world/

Wiggers, K. (2023, March 28). *VCS continue to pour dollars into Generative AI.* TechCrunch. https://techcrunch.com/2023/03/28/generative-ai-venture-capital/

Chapter 10 – Workforce Transformation

593 Digital publisher CEIT. (n.d.-a). https://www.593dp.com/index.php/593_Digital_Publisher

Aidude. (2023). *AI tools for business.* Find the Best AI Tools to Improve Your Business. https://aidude.info/categories/business

Amir, A. (2023, May 31). *The future of work: Embracing automation and the gig economy.* Ek ajanbee. https://ekajanbee.in/2023/05/the-future-of-work-embracing-automation-and-the-gig-economy.html

Basu, S. (2023, May). *Responsible AI.* Oracle's Guide to Ethical Considerations in AI Development and Deployment. https://docs.oracle.com/en-us/iaas/Content/Resources/Assets/whitepapers/responsible-ai-oracle.pdf

Can ai replace humans or can not? 10 great arguments revealed - noowai. NoowAI.com - Ask me everything...! (2023, May 30). https://noowai.com/can-ai-replace-humans/

Dye, E. (2023, May 11). *Esteban dye on linkedin: What is it like to be an AI?.* Esteban Dye on LinkedIn: What is it like to be an AI? https://www.linkedin.com/posts/estebandye_what-is-it-like-to-be-an-ai-activity-7062219142426181632-xk7t

Exploring the power of ai: Redefining work in the age of Artificial Intelligence. Learnow. (n.d.). https://www.learnow.live/blog/what-does-it-mean-to-work-in-the-age-of-ai

The future of work and the gig economy in the Digital age. SME Cloud. (n.d.). https://smecloud.my/the-future-of-work-and-the-gig-economy-in-the-digital-age/

Gandzeichuk, I. (2023, April 25). *Council post: How ai can transform the software engineering process.* Forbes. https://www.forbes.com/sites/forbestechcouncil/2023/04/24/how-ai-can-transform-the-software-engineering-process/

Gandzeychuk, I. (2023, April). *How AI can transform the software engineering process.* LinkedIn. https://www.linkedin.com/pulse/how-ai-can-transform-software-engineering-process-ilya-gandzeychuk

Ibia. (n.d.). https://www.ibia.org/email_ibia.phpx?id=3803

The impact of AI on developer productivity: Evidence from github copilot. Papers With Code. (n.d.). https://cs.paperswithcode.com/paper/the-impact-of-ai-on-developer-productivity

The impact of artificial intelligence on the future of workforces in … (n.d.-b). https://www.whitehouse.gov/wp-content/uploads/2022/12/TTC-EC-CEA-AI-Report-12052022-1.pdf

LO Blog. (2023, April 1). *The Future of Artificial Intelligence (AI).* LEARNING. https://www.learningoffical.com/2023/03/The-Future-of-Artificial-Intelligence.html

MacKenzie, K. (2023, May 4). *How AI can enhance human skills and collaboration at work.* Recruiting Resources: How to Recruit and Hire Better. https://resources.workable.com/stories-and-insights/how-ai-can-enhance-human-skills-and-collaboration-at-work

Pat. (2023, April 7). *The Future of Data Entry: The advantages of Ai Assistance.* Data Entry Company. https://dataentrycompany.com/the-future-of-data-entry-the-advantages-of-ai-assistance/

Peng, S., Kalliamvakou, E., Cihon, P., & Demirer, M. (n.d.-a). *The impact of AI on developer productivity: Evidence from github copilot.* – arXiv Vanity. https://www.arxiv-vanity.com/papers/2302.06590/

Peng, S., Kalliamvakou, E., Cihon, P., & Demirer, M. (n.d.-b). *The impact of AI on developer productivity: Evidence from github copilot.* NASA/ADS. https://ui.adsabs.harvard.edu/abs/2023arXiv230206590P/abstract

Pros and cons of artificial intelligence in the workplace 2023. Ablison. (2023, March 3). https://www.ablison.com/pros-and-cons-of-artificial-intelligence-in-the-workplace/

Revell, E. (2023, May 8). *Uber seeks patent to "pre-match" riders and drivers using AI.* Fox Business. https://www.foxbusiness.com/technology/uber-seeks-patent-pre-match-riders-drivers-using-ai

School, M. (2023, May 29). *Putting AI and automation to work for you: Harnessing the power of technology for efficiency and…* Medium. https://medium.com/@moringa_school/putting-ai-and-automation-to-work-for-you-harnessing-the-power-of-technology-for-efficiency-and-33a349b0f3ce?source=tag_page---------37-84--------------------6e95616d_3865_4c8d_913e_364ad3a5277c-------17

Spencer, M. (2023, April 28). *📖 the impact of A.I. on developer productivity and jobs.* 📖 The Impact of A.I. on Developer Productivity and Jobs. https://aisupremacy.substack.com/p/the-impact-of-ai-on-developer-productivity

Streamline your workflow: A beginner's Guide to Business Automation. Strictly Savvy | Streamline Your Workflow: A Beginner's Guide to Business Automation. (2023, April 5). https://www.strictlysavvy.co.nz/blog/post/102866/streamline-your-workflow-a-beginners-guide-to-business-automation/

Syed, W. (2023, May 20). *The rise of the Gig Economy: Shaping the future of work.* Technomics Global. https://technomicsglobal.com/2023/05/20/the-rise-of-the-gig-economy-shaping-the-future-of-work/

team, E. (2023, April 21). *The power of AI in policymaking: Advancements, applications, and challenges.* Bharti Institute of Public Policy. https://blogs.isb.edu/bhartiinstitute/2023/03/07/the-power-of-ai-in-policymaking-advancements-applications-and-challenges/

Walid Hariri arXiv:2304.02017v5 [cs.CL] 12 Apr 2023. (n.d.-c). https://arxiv.org/pdf/2304.02017.pdf

Weiser, B. (2023, May 27). *Here's what happens when your lawyer uses Chatgpt.* The New York Times. https://www.nytimes.com/2023/05/27/nyregion/avianca-airline-lawsuit-chatgpt.html

What is Artificial Intelligence and machine learning?. Tactical Rabbit. (2023, March 27). https://tacticalrabbit.com/services/artificial-intelligence-capabilities/what-is-artificial-intelligence-and-machine-learning/

Written BySingle Grain Team Single Grain is a full-service digital marketing agency that helps great companies grow. Our team of experts is passionate about helping brands expand with SEO. (2023, May 22). *How to implement AI in your business: A step-by-step guide.* Single Grain. https://www.singlegrain.com/blog/how-to-implement-ai-in-my-business/

Chapter 11 - Healthcare

Baclic, O., Tunis, M., Young, K., Doan, C., Swerdfeger, H., & Schonfeld, J. (2020, June 4). *Challenges and opportunities for public health made possible by advances in Natural Language Processing.* Canada communicable disease report = Releve des maladies transmissibles au Canada. https://www.ncbi.nlm.nih.gov/pmc/articles/PMC7343054/

Bajwa, J., Munir, U., Nori, A., & Williams, B. (2021, July). *Artificial Intelligence in healthcare: Transforming the practice of medicine.* Future healthcare journal. https://www.ncbi.nlm.nih.gov/pmc/articles/PMC8285156/

Bateman, K. (2021, December). *4 ways artificial intelligence is improving mental health therapy.* World Economic Forum. https://www.weforum.org/agenda/2021/12/ai-mental-health-cbt-therapy/

Bryant, M. (2019, March 29). *How AI and Machine Learning Are Changing prosthetics.* MedTech Dive. https://www.medtechdive.com/news/how-ai-and-machine-learning-are-changing-prosthetics/550788/

Chen, A. (2018, July 26). *IBM's Watson gave unsafe recommendations for treating cancer.* The Verge. https://www.theverge.com/2018/7/26/17619382/ibms-watson-cancer-ai-healthcare-science

Corbyn, Z. (2021, June 3). *The future of elder care is here – and it's artificial intelligence.* The Guardian. https://www.theguardian.com/us-news/2021/jun/03/elder-care-artificial-intelligence-software

Fatpos Global Pvt. Ltd. (n.d.). *AI for Drug Discovery Market by offering.* AI for Drug Discovery Market By Offering ;By Technology ; By Drug Type ; By Application ; By End-User and Region – Analysis of Market Size, Share & Trends for 2018 – 2020 and Forecasts to 2030. https://www.reportlinker.com/p06191597/AI-for-Drug-Discovery-Market-By-Offering-By-Technology-By-Drug-Type-By-Application-By-End-User-and-Region-Analysis-of-Market-Size-Share-Trends-for-and-Forecasts-to.html?utm_source=GNW

Hall, R. (2023, June 9). *Unleashing the power of Artificial Intelligence: Transforming Industries with ai.* LinkedIn. https://www.linkedin.com/pulse/unleashing-power-artificial-intelligence-transforming-rob-hall

Johnson, K. B., Wei, W.-Q., Weeraratne, D., Frisse, M. E., Misulis, K., Rhee, K., Zhao, J., & Snowdon, J. L. (2021, January). *Precision Medicine, AI, and the future of Personalized Health Care.* Clinical and translational science. https://www.ncbi.nlm.nih.gov/pmc/articles/PMC7877825/

Jones, S. (2023, June 5). *How can ai be used to address public health challenges?.* Webmedy. https://webmedy.com/blog/ai-public-health/

Kitson, S. (2023, March 29). *Revolutionising medical imaging with AI and Big Data Analytics*. Open Medscience. https://openmedscience.com/revolutionising-medical-imaging-with-ai-and-big-data-analytics/

Kizen. (2023, June 14). *How ai in healthcare transforms patient journeys & experiences*. Kizen. https://kizen.com/content/ai-ml/ai-in-healthcare/

Kwak, L., & Bai, H. (2023). The role of Federated Learning Models in medical imaging. *Radiology: Artificial Intelligence, 5*(3). https://doi.org/10.1148/ryai.230136

Lomis, K., Jeffries, P., Palatta, A., Sage, M., Sheikh, J., Sheperis, C., & Whelan, A. (2021, September 8). *Artificial Intelligence for Health Professions Educators*. NAM perspectives. https://www.ncbi.nlm.nih.gov/pmc/articles/PMC8654471/

Mahajan, P. Suresh. (2022). *Artificial Intelligence in Healthcare: AI, Machine Learning, and deep and intelligent medicine simplified for everyone*. Amazon. https://www.amazon.com/Artificial-Intelligence-Healthcare-Intelligent-Simplified/dp/1954612028

Mahajan, R. (2023). *Quantum Care: A deep dive into AI for Health Delivery and Research*. Amazon. https://www.amazon.com/Quantum-Care-Health-Delivery-Research/dp/1642255548

Meyer, M. (2021, June 15). *AI and IOT Convergence for smart health*. IEEE Access. https://ieeeaccess.ieee.org/closed-special-sections/ai-and-iot-convergence-for-smart-health/

Mills, T. (2022, February 18). *Council post: AI for health and hope: How machine learning is being used in hospitals*. Forbes. https://www.forbes.com/sites/forbestechcouncil/2022/02/16/ai-for-health-and-hope-how-machine-learning-is-being-used-in-hospitals/

Muller, R. (2021, March). *Is artificial intelligence the future of mental health?*. Psychology Today. https://www.psychologytoday.com/us/blog/talking-about-trauma/202103/is-artificial-intelligence-the-future-mental-health

Nathan, J. (2023, February 16). *Council post: Four Ways Artificial Intelligence can benefit robotic surgery*. Forbes. https://www.forbes.com/sites/forbestechcouncil/2023/02/15/four-ways-artificial-intelligence-can-benefit-robotic-surgery/

O'Leary, L. (2022, January 31). *How IBM's Watson went from the future of health care to sold off for parts*. Slate Magazine. https://slate.com/technology/2022/01/ibm-watson-health-failure-artificial-intelligence.html

Paul, D., Sanap, G., Shenoy, S., Kalyane, D., Kalia, K., & Tekade, R. K. (2021, January). *Artificial Intelligence in drug discovery and development*. Drug discovery today. https://www.ncbi.nlm.nih.gov/pmc/articles/PMC7577280/

Payne, A. (2020, September 22). *The role of AI in Advancing Personalized Healthcare*. TechRadar. https://www.techradar.com/news/the-role-of-ai-in-advancing-personalized-healthcare

Philips. (2021). How AI can enhance the human experience in healthcare - philips. https://www.philips.com/c-dam/corporate/newscenter/global/standard/resources/healthcare/2021/enhance-human-experience/philips-ai-position-paper.pdf

Philips. (2022, November 24). *10 real-world examples of AI in Healthcare*. Philips. https://www.philips.com/a-w/about/news/archive/features/2022/20221124-10-real-world-examples-of-ai-in-healthcare.html

Ram, A. (2018, February 4). *Deepmind develops AI to diagnose eye diseases.* Subscribe to read | Financial Times. https://www.ft.com/content/84fcc16c-0787-11e8-9650-9c0ad2d7c5b5

Roller, J. (2023, May 13). *Revolutionizing Healthcare: The impact of AI on medical diagnoses and treatment decisions.* IEEE Computer Society. https://www.computer.org/publications/tech-news/community-voices/ai-impact-on-medical-diagnosis-treatment

Schork, N. J. (2019). *Artificial Intelligence and personalized medicine.* Cancer treatment and research. https://www.ncbi.nlm.nih.gov/pmc/articles/PMC7580505/

Tamini, K. H. (n.d.). *Unraveling the AI and human relationship: A modern egg and chicken dilemma.* LinkedIn. https://www.linkedin.com/pulse/unraveling-ai-human-relationship-modern-egg-chicken-heidartamini/

USAID. (2022, May 26). *Artificial Intelligence in global health: Defining a collective path forward: Global Health.* U.S. Agency for International Development. https://www.usaid.gov/cii/ai-in-global-health

Chapter 12 – Marketing and Sales

Ai in sales: Benefits of using AI for sales: Chatfuel Blog. Chatfuel. (n.d.). https://chatfuel.com/blog/ai-sales

Ai lead generation: Maximizing Roi with Artificial Intelligence. Outgrow. (2023, May 30). https://outgrow.co/blog/ai-lead-generation

Akanni, D. (n.d.). The future of marketing is undoubtedly AI-driven, and those who embrace it will reap the rewards of this powerful duo. LinkedIn. https://www.linkedin.com/pulse/future-marketing-undoubtedly-ai-driven-those-who-embrace-akanni/

Benefits of chatbots for customers - linkedin. (n.d.). https://www.linkedin.com/pulse/benefits-chatbots-customers-it-khabir

Breakthrough, B. (2023, May 19). AI-powered customer service: How chatbots and virtual assistants are transforming business... Medium. https://ai.plainenglish.io/ai-powered-customer-service-how-chatbots-and-virtual-assistants-are-transforming-business-9754cc63204d

Chatbot – arkatiss: Custom software development. Arkatiss custom software development. (n.d.). https://eagle.arkatiss.com/chatbot/

ChatGPT API based Chatbot Free Tool Online. GPTTOLDCOM. (n.d.). https://www.gpttold.com/

Craig, S. (2023, April 20). Revolutionizing sales with Ai Chatbots. Elite Chat. https://elitechat.io/revolutionizing-sales-with-ai-chatbots/

Customer engagement in 2023 ValueFirst, a twilio company. ValueFirst, a Twilio companyCommunication Platform for SMS APIs, Voice APIs, Video APIs - ValueFirst, a Twilio company. (n.d.). https://www.vfirst.com/post/customer-engagement-in-2023

Davenport, T. H., Guha, A., & Grewal, D. (2021). How to design an AI marketing strategy. Harvard Business Review. https://hbr.org/2021/07/how-to-design-an-ai-marketing-strategy

Davison, A. (2023, May 22). Responsible regulation of generative AI needed - it-online. IT. https://it-online.co.za/2023/05/22/responsible-regulation-of-generative-ai-needed/

Dilmegani, C. (2023, March 6). Dynamic pricing algorithms in 2023: Top 3 models. AIMultiple. https://research.aimultiple.com/dynamic-pricing-algorithm/

Editor. (2023, May 6). 10 proven lead nurturing tactics for successful lead campaign. IT Software Tips. https://itsoftwaretips.com/10-proven-lead-nurturing-tactics-for-successful-lead-campaign/

Egol, M. (n.d.). CHATGPT answers ten key questions about how AI will impact CX. LinkedIn. https://www.linkedin.com/pulse/chatgpt-answers-ten-key-questions-how-ai-impact-cx-matthew-egol/

FCR - ContactCenterWorld.com blog. ContactCenterWorld.com. (n.d.). https://www.contactcenterworld.com/company/blog/fcr/

Frąckiewicz, M. (2023a, April 25). AI in retail: Enhancing customer experience and streamlining operations. TS2 SPACE. https://ts2.space/en/ai-in-retail-enhancing-customer-experience-and-streamlining-operations/

Frąckiewicz, M. (2023b, May 22). The role of Artificial Intelligence in predictive analytics. TS2 SPACE. https://ts2.space/en/the-role-of-artificial-intelligence-in-predictive-analytics/

Frąckiewicz, M. (2023c, May 23). The impact of AI on marketing: Personalization, prediction, and performance. TS2 SPACE. https://ts2.space/en/the-impact-of-ai-on-marketing-personalization-prediction-and-performance/

Frąckiewicz, M. (2023d, May 26). Embracing AI in e-commerce: A strategic move towards Digital Transformation. TS2 SPACE. https://ts2.space/en/embracing-ai-in-e-commerce-a-strategic-move-towards-digital-transformation/

The Future of Artificial Intelligence: How Ai is Shaping Industries. Education. (n.d.). https://vocal.media/education/the-future-of-artificial-intelligence-how-ai-is-shaping-industries

Gkikas, D. C., & Theodoridis, P. K. (2021, October). Ai in consumer behavior. SpringerLink. https://link.springer.com/chapter/10.1007/978-3-030-80571-5_10

Gupta, A. (2022, April 21). Council post: 14 smart ways marketers and advertisers can leverage AI in 2022. Forbes. https://www.forbes.com/sites/forbesagencycouncil/2021/12/15/14-smart-ways-marketers-and-advertisers-can-leverage-ai-in-2022/

Hernandez, A. (2023, April 10). My Honest Review of writesonic ai writing. Don't Do It Yourself. https://ddiy.co/writesonic-ai-review/

Hyvärinen, J. (2023, June 1). The future of Personalization: How Ai and Machine Learning Are Transforming Digital Marketing. Ranktracker Blog RSS. https://www.ranktracker.com/blog/the-future-of-personalization-how-ai-and-machine-learning-are-transforming-digital-marketing/

Inge, C. (2022). Marketing metrics: Leverage analytics and data to optimize marketing strategies. Amazon. https://www.amazon.com/Marketing-Metrics-Leverage-Analytics-Strategies/dp/1398606596

McBeth, G. (2023, February 16). The future of AI for sales (and how to prepare for it). Sales Hacker. https://www.saleshacker.com/ai-for-sales/

Payani, A. (2023, March 9). Council post: Embracing the future: How ai is Revolutionizing Marketing and Sales. Forbes. https://www.forbes.com/sites/forbesbusinesscouncil/2023/03/08/embracing-the-future-how-ai-is-revolutionizing-marketing-and-sales/

Paylocity. (2021, April 28). Paylocity collaborates with Deloitte and Launches Modern Workforce Research. PR Newswire: press release distribution, targeting, monitoring and marketing. https://www.prnewswire.com/news-releases/paylocity-collaborates-with-deloitte-and-launches-modern-workforce-research-301278409.html?tc=eml_cleartime

Pega Systems. (2023, May 11). Real-time, Omni-channel AI. Pega. https://www.pega.com/insights/resources/real-time-omni-channel-ai

the power of AI (Artificial Intelligence) and machine learning is reshaping various industries, including marketing. T. article explores the. (2023, May 26). The impact of AI and machine learning on Digital Marketing Analytics in 2023. Digittal Leaf. https://www.digittalleaf.com/2023/05/ai-analytics-marketing.html

Saxena, P. (2023, May 19). Generative AI unleashed: Revolutionizing business innovation in the Digital age. Medium. https://levelup.gitconnected.com/generative-ai-unleashed-revolutionizing-business-innovation-in-the-digital-age-899d02c94a29

Shpitula, N. (2023, May 24). Ai in e-commerce industry: Importance, benefits, and use cases. Soloway. https://soloway.tech/blog/ai-in-ecommerce-benefits-and-examples/

Singh, J. (2023, June 13). Nvidia-backed platform that turns text into a.i.-generated avatars boosts valuation to $1 billion. MSN. https://www.msn.com/en-us/money/companies/nvidia-backed-platform-that-turns-text-into-a-i-generated-avatars-boosts-valuation-to-1-billion/ar-AA1cu4Vd

Singh, R. (2023, May 3). Artificial Intelligence & Machine Learning in website development. Website Monitoring Server Monitoring Blog. https://www.websitepulse.com/blog/artificial-intelligence-machine-learning-website-development

Sinha, P., Shastri, A., & Lorimer, S. E. (2023, March 31). How generative AI will change sales. Harvard Business Review. https://hbr.org/2023/03/how-generative-ai-will-change-sales

Sowery, K. (n.d.). The era of AI: How is Digital Marketing being transformed?. Startups Magazine. https://startupsmagazine.co.uk/article-era-ai-how-digital-marketing-being-transformed

Starita, L. (2022, June 16). AI in marketing: Benefits, use cases, and examples. Persado. https://www.persado.com/articles/ai-marketing/

Unleashing the power of ai: Transforming business strategy for Success. LinkedIn. (n.d.). https://www.linkedin.com/pulse/unleashing-power-ai-transforming-business-strategy-success/

Using customer segmentation to improve retention rates. Markettailor. (n.d.). https://www.markettailor.io/blog/customer-segmentation-to-improve-retention-rates

Venkatesan, R., & Lecinski, J. (2021). The AI marketing canvas: A five-stage road map to Implementing Artificial Intelligence in marketing. Stanford Business Books, an imprint of Stanford University Press.

Vince. (2023, May 29). Artificial Intelligence Demystified: Unlocking growth and efficiency through SMART Software Solutions. Hardball Reviews.

https://hardballreviews.com/artificial-intelligence-demystified-unlocking-growth-and-efficiency-through-smart-software-solutions/

Wisneski, C. (2021, June 3). Complete guide to using AI predictive lead scoring in sales. Akkio. https://www.akkio.com/post/complete-guide-to-using-ai-predictive-lead-scoring-in-sales

Zaki, M., McColl-Kennedy, J. R., & Neely, A. (2021, May 4). Using AI to track how customers feel - in real time. Harvard Business Review. https://hbr.org/2021/05/using-ai-to-track-how-customers-feel-in-real-time

Zupko, A. (n.d.). The power of AI: Revolutionizing Business Operations and driving growth. https://www.linkedin.com/pulse/power-ai-revolutionizing-business-operations-driving-growth-zupko

Chapter 13 – Financial Services

Baheti, P. (2022, August 31). *9 innovative use cases of AI in finance [+pros & cons]*. V7. https://www.v7labs.com/blog/ai-in-finance

Falk, R. (2023, June 22). *Blogs*. Amazon. https://aws.amazon.com/blogs/industries/the-next-frontier-generative-ai-for-financial-services/

Haas, C., & Gilmore, A. (2023, March 30). Introducing BloombergGPT, Bloomberg's 50-billion parameter large language model, purpose-built from scratch for Finance | Press | Bloomberg LP. Bloomberg.com. https://www.bloomberg.com/company/press/bloomberggpt-50-billion-parameter-llm-tuned-finance/

IBIS. (2023, May 31). *Industry market research, reports, and Statistics*. IBISWorld. https://www.ibisworld.com/united-states/market-research-reports/financial-data-service-providers-industry/

McKendrick, J., & Margaris, S. (2023, January 16). *The coming democratization of financial services, thanks to Artificial Intelligence*. Forbes. https://www.forbes.com/sites/joemckendrick/2023/01/14/the-coming-democratization-of-financial-services-thanks-to-ai/?sh=5731c170582b

Patel, B. (2018, March 22). *How ai is transforming trade settlements*. Finextra Research. https://www.finextra.com/blogposting/15175/how-ai-is-transforming-trade-settlements#:~:text=AI%20is%20capable%20of%20detecting,of%20investment%20by%20financial%20institutions.

Savi, R., Shen, J., & Tsaig, Y. (2023, June 15). *How ai is transforming investing*. BlackRock. https://www.blackrock.com/us/individual/insights/ai-investing

Chapter 14 - Education

Amesite LMS. (2023, April 3). *Blogs*. Amesite " Professional Certificates: How AI Is Reshaping Access to Learning. https://amesite.com/blogs/professional-certificates-how-ai-is-reshaping-access-to-learning/

Carnegie Learning. (2023). *Mathia*. Carnegie Learning. https://www.carnegielearning.com/solutions/math/mathia/

Dickson, B. (2017, November 20). *How artificial intelligence is shaping the future of Education.* PCMAG. https://www.pcmag.com/news/how-artificial-intelligence-is-shaping-the-future-of-education

Dimeo, J. (2017, July 18). *Georgia state improves student outcomes with data.* Inside Higher Ed | Higher Education News, Events and Jobs. https://www.insidehighered.com/digital-learning/article/2017/07/19/georgia-state-improves-student-outcomes-data

Frąckiewicz, M. (2023, May 25). *Artificial Intelligence: The Future of Personalized Learning.* TS2 SPACE. https://ts2.space/en/artificial-intelligence-the-future-of-personalized-learning/

Greene-Haroer, R. T. (2023, April 27). *The Pros and cons of using AI in learning: Is CHATGPT helping or hindering learning outcomes?.* eLearning Industry. https://elearningindustry.com/pros-and-cons-of-using-ai-in-learning-chatgpt-helping-or-hindering-learning-outcomes

Gülen, K. (2023, February 8). *How AI improves education with personalized learning at scale and other new capabilities.* Dataconomy. https://dataconomy.com/2023/02/03/artificial-intelligence-in-education/

Harve, A. (2023, June 6). *Transforming education with AI: Innovations in curriculum design.* Hurix Digital. https://www.hurix.com/transforming-education-with-ai-innovations-in-curriculum-design/

K, M. (2023, June 6). *How AI is personalizing education for every student.* eLearning Industry. https://elearningindustry.com/how-ai-is-personalizing-education-for-every-student

Louder, J. (2023, March 29). *The future of AI in education: How will it transform learning?.* Anthology. https://www.anthology.com/blog/the-future-of-ai-in-education-how-will-it-transform-learning

Marr, B. (2020, October 16). *The amazing ways duolingo is using artificial intelligence to deliver free language learning.* Forbes. https://www.forbes.com/sites/bernardmarr/2020/10/16/the-amazing-ways-duolingo-is-using-artificial-intelligence-to-deliver-free-language-learning/

Pearson. (2023). *Large Scale Educational Assessment, scoring, and reporting.* Pearson. https://www.pearsonassessments.com/large-scale-assessments/k-12-large-scale-assessments/automated-scoring.html

Wang, T., Lund, B. D., Marengo, A., Pagano, A., Mannuru, N. R., Teel, Z. A., & Pange, J. (2023, May 31). Exploring the potential impact of Artificial Intelligence (AI) on international students in Higher Education: Generative AI, Chatbots, analytics, and International Student Success. MDPI. https://www.mdpi.com/2076-3417/13/11/6716

Zdravkova, K., Krasniqi, V., Dalipi, F., & Ferati, M. (2022, September 27). *Cutting-edge communication and learning assistive technologies for disabled children: An artificial intelligence perspective.* Frontiers. https://www.frontiersin.org/articles/10.3389/frai.2022.970430/full

Chapter 15 – Authenticity

Bolden, D., Bellefonds, N. de, & Duranton, S. (2023). *Artificial Intelligence.* BCG Global. https://www.bcg.com/capabilities/digital-technology-data/artificial-intelligence

Brodsky, A. (2022, January 20). *Communicating authentically in a virtual world.* Harvard Business Review. https://hbr.org/2022/01/communicating-authentically-in-a-virtual-world

Buhr, H., Funk, R. J., & Owen-Smith, J. (2021). The authenticity premium: Balancing conformity and innovation in high technology industries. *Research Policy, 50*(1), 104085. https://doi.org/10.1016/j.respol.2020.104085

Cerejo, L. (2023, April 3). *Beyond algorithms: Skills of designers that ai can't replicate.* Smashing Magazine. https://www.smashingmagazine.com/2023/04/skills-designers-ai-cant-replicate/

Cerejo, L., About The AuthorLyndon Cerejo is a UX Design leader with over twenty-five years of hands-on experience helping companies design usable and engaging experiences for their ...More aboutLyndon ↪, & Author, A. T. (2023, April 3). *Beyond algorithms: Skills of designers that ai can't replicate.* Smashing Magazine. https://www.smashingmagazine.com/2023/04/skills-designers-ai-cant-replicate/

Dey, B. (2023, April 27). *Is AI-generated content – ethical, authentic and efficient?.* Writingo. https://blogs.writingo.ai/is-ai-generated-content-ethical-authentic-and-efficient/

Gallagher, L. (2018). The Airbnb Story: How Three Ordinary Guys disrupted an industry, made billions... and created plenty of controversy. Mariner Books.

Helmore, E. (2022, September 4). *The US Artisan Revolution: How the simple life came in from the margins.* The Guardian. https://www.theguardian.com/focus/2022/sep/04/the-us-artisan-revolution-how-the-simple-life-came-in-from-the-margins

Ibrahim&Co. (2023, May 29). *Valuing the human touch: Ai, authenticity, and the future of original content.* Medium. https://medium.com/@ibrahimandcompany/valuing-the-human-touch-ai-authenticity-and-the-future-of-original-content-7b44e9b5d66d

Lee, V. R. (2023a, May 5). Can authenticity exist in the age of AI? https://www.fastcompany.com/90891866/authenticity-age-of-ai

Lee, V. R. (2023b, May 19). *Generative AI is forcing people to rethink what it means to be authentic.* The Conversation. https://theconversation.com/generative-ai-is-forcing-people-to-rethink-what-it-means-to-be-authentic-204347

Ordoñez, L. (2018, February). *Succcessful ecommerce case: The history of Zappos.* Oleoshop. https://www.oleoshop.com/en/blog/succcessful-ecommerce-case-the-history-of-zappos

OutCrowd. (2023, May 20). *The power of storytelling in digital marketing.* Boost Framer website template. https://outcrowdx.com/blog/the-power-of-storytelling-in-digital-marketing

PYMNTS.com. (2022, August 9). *Allbirds' omnichannel evolution from D2C Pure-play can't happen fast enough.* Pymnts.com. https://www.pymnts.com/news/retail/2022/allbirds-omnichannel-evolution-from-d2c-pure-play-cannot-happen-fast-enough/

Randieri, C. (2023, March 27). *Council post: Can ai replace human curiosity?.* Forbes. https://www.forbes.com/sites/forbestechcouncil/2023/03/22/can-ai-replace-human-curiosity

Sharma, S. (2023, May 20). *Shashank Sharma on linkedin: #prompted.* Shashank Sharma on LinkedIn: #prompted. https://www.linkedin.com/posts/shashank-sharma-aa85a68_prompted-activity-7065691785401946112-VIpD/

Steiner, C. (2021, April 19). *How to build a direct-to-consumer brand: Data, authenticity, CX and Gore-Tex.* Forbes. https://www.forbes.com/sites/christophersteiner/2021/04/08/how-to-build-a-direct-to-consumer-brand-data-authenticity-cx-and-gore-tex/?sh=56f124274f8c

Tran, A. (2022, September 2). *Unrolling glossier: Building a beauty and Wellness Ecommerce Empire.* AdRoll. https://www.adroll.com/blog/unrolling-glossier-building-a-beauty-and-wellness-ecommerce-empire

Whitfield, D. (2021, March 19). *Council post: How to build brand authenticity with artificial intelligence.* Forbes. https://www.forbes.com/sites/forbestechcouncil/2021/03/19/how-to-build-brand-authenticity-with-artificial-intelligence/

Chapter 16 – Future of AI

Al-Ghourabi, A. (2023, May 16). *The democratization of AI: The rise of solopreneurs and the evolution of business.* LinkedIn. https://www.linkedin.com/pulse/democratization-ai-rise-solopreneurs-evolution-ad-al-ghourabi/

Alteryx. (2023, May 16). *The Essential Guide to Explainable AI (XAI).* Alteryx. https://www.alteryx.com/resources/whitepaper/essential-guide-to-explainable-ai

Amir, M. (2023, February 13). *The future of AI and its impact on society.* Amir Blogs. https://amirblogs.com/future-of-ai-and-its-impact-on-society/

Defense Counsel Journal. IADC. (n.d.). https://www.iadclaw.org/defensecounseljournal/this-internet-thing-is-great-isnt-it-hal-product-liability-in-the-next-100-years

The ethics of data science, Balancing Innovation and responsibility. Futurism. (n.d.). https://vocal.media/futurism/the-ethics-of-data-science-balancing-innovation-and-responsibility

European Parliment. (2022, January). *Artificial Intelligence Act: Think tank: European parliament.* Think Tank | European Parliament. https://www.europarl.europa.eu/thinktank/en/document/EPRS_BRI%282021%29296 98792

Fortune Insights. (2023, April). *Artificial Intelligence [AI] market size, share & forecast, 2030.* Artificial Intelligence [AI] Market Size, Share & Forecast, 2030. https://www.fortunebusinessinsights.com/industry-reports/artificial-intelligence-market-100114

Garg, S. (2023, May 15). *What is autogpt? A comprehensive guide & CHATGPT comparison.* The Writesonic Blog - Making Content Your Superpower. https://writesonic.com/blog/what-is-autogpt/

Hua, H., Li, Y., Wang, T., Dong, N., Li, W., & Cao, J. (2023, January 1). *Edge computing with Artificial Intelligence: A machine learning perspective.* ACM Computing Surveys. https://dl.acm.org/doi/full/10.1145/3555802

Kang, C. (2023, May 16). *OpenAI's Sam Altman urges A.I. Regulation in Senate hearing.* The New York Times. https://www.nytimes.com/2023/05/16/technology/openai-altman-artificial-intelligence-regulation.html

Karakas, F. (2023, February 5). *I sought the help of artificial intelligence to understand singularity.* Medium. https://medium.com/predict/i-sought-the-help-of-artificial-intelligence-to-understand-singularity-96ed54ec5720

McKendrick, J. (2023, May 30). *Artificial Intelligence's higher value: Spurring new managerial thinking.* Forbes. https://www.forbes.com/sites/joemckendrick/2023/05/20/artificial-intelligence-will-eliminate-many-tasks-but-also-create-new-higher-level-ones

Oliver, T. (2023, April 17). *How AI is boosting profits through revolutionary data analysis techniques.* INQUIRER.net. https://business.inquirer.net/396219/how-ai-is-boosting-profits-through-revolutionary-data-analysis-techniques

Sharma, S. (2023, May 20). *Shashank Sharma on linkedin: #prompted.* Shashank Sharma on LinkedIn: #prompted. https://www.linkedin.com/posts/shashank-sharma-aa85a68_prompted-activity-7065691785401946112-VIpD/

Shukla, M. (2019, November 7). *Council post: The Democratization of Technology.* Forbes. https://www.forbes.com/sites/forbestechcouncil/2019/11/07/the-democratization-of-technology/?sh=51ea00f83796

Stern, S. (n.d.). *The future of nonprofit work: How ai tools are changing the game.* LinkedIn. https://www.linkedin.com/pulse/future-nonprofit-work-how-ai-tools-changing-game-sam-stern

Thiel, T. (2022, January 6). *Artificial Intelligence and democracy: Heinrich-Böll-Stiftung: Tel Aviv - israel.* Heinrich-Böll-Stiftung. https://il.boell.org/en/2022/01/06/artificial-intelligence-and-democracy

Thiel, T., Gregorio, G. D., Feldman, D. K.-D., & Bürklen, A. (2023, February 6). *Artificial Intelligence and democracy.* IPPI. https://www.ippi.org.il/artificial-intelligence-and-democracy/

Wynsberghe, A. van. (2021, February 26). *Sustainable AI: AI for Sustainability and the sustainability of Ai - Ai and Ethics.* SpringerLink. https://link.springer.com/article/10.1007/s43681-021-00043-6

INDEX